THE
O.
5

· THE DIAMOND MATCH CO. IN THE HIGH SIERRA ·

Matches, Flumes, and Rails

First No. 6 while it was in service on the logging railroad. (Mrs. Florence Styles)

by

Kent Stephens

TRANS-ANGLO BOOKS

Library of Congress Cataloging in Publication Data

Stephens, Kent.
Matches, flumes, and rails.

Bibliography: p.
Includes index.
1. Diamond Gardner Corporation. I. Title.
HD9999.M24D77 338.7'66'25097944 77-76014
ISBN 0-87046-030-7

MATCHES, FLUMES & RAILS
The Diamond Match Company in the High Sierra

by KENT STEPHENS

FIRST EDITION

Library of Congress Catalog Card Number: 77-76014

ISBN: 0-87046-030-7

To Joanne, Kristy, and Johnnie

Dust Jacket: Gleaming 2-8-0 No. 4 ready to leave Stirling City with a one-car special. (D. I. Corp. collection)

End Papers: No. 5 and crew pose for the photographer on the Kimshew line. (George Kahn photo, D. I. Corp. collection)

Frontispiece: A railroad construction crew erects a trestle on the Kimshew logging railroad. (George Kahn photo, D. I. Corp. collection)

BOOK DESIGN: HANK JOHNSTON

Editorial Director: Spencer Crump

Printed and Bound in the United States of America

•

Published by Trans-Anglo Books,
P.O. Box 38, Corona del Mar, California 92625

CONTENTS

MAP OF THE

STANDARD AND NARROW GAUGE

LOGGING RAILROADS

OF

THE DIAMOND MATCH COMPANY

compiled by KENT STEPHENS drawn by DAVID W. BRAUN

1973

GRAPHIC SCALE ~ 1 mile

Acknowledgments

The first work on this book actually started long before it was planned. I received approval from Mr. William H. Hutchinson and the Diamond International Corporation in late 1962 to utilize research material in the corporation's California Lumber Division archives to write an article on the Butte County Railroad. This research material had been developed some five years earlier by Mr. Hutchinson when he wrote two books on the history of the corporation's California lumbering activities. Only one of these received wide public distribution. This was *California Heritage: A History of Northern California Lumbering,* an excellent history of Sierra Lumber Company and its predecessors. Almost the only available work on lumbering in northeastern California, it was out of print until republished in 1975 by the Forest History Society.

The second book by Mr. Hutchinson was on the Diamond Match Company itself. Titled *The California Investment of The Diamond Match Company,* it received limited public distribution by the corporation and has long been out of print.

Mr. Hutchinson, familiarly known as "Old Hutch" and a leading Western historian, had a specified length of time to complete his research on all aspects of Diamond's activities in California. I have extended the research on Diamond's railroads, and with the approval of Mr. Hutchinson and the corporation, have utilized the archival material to complete the history of the general aspects of the corporation. My research has been greatly eased by the existence of the archives and I would like to express my appreciation to Mr. Hutchinson and the Diamond International Corporation for their use.

Access to the archives was freely provided by C. L. Millman, Larry Fedler, and later by Mrs. Thelma Beasley. Other personnel who have been helpful include Mel Jones, former Vice President of the California Lumber Division; Richard Roseberry; Fred Raymond; Barry Ford; W. N. Dennison; Leon N. Brown of Diamond's Springfield (Massachusetts) plant and Jerry F. Pervinich of the New York office.

Three retired employees closely connected with the California Lumber Division's logging operations and the logging railroad deserve special thanks for the help they extended. Richard A. "Dick" Colgan, Jr. started his career with Diamond in California during the early 1920's when he joined the corporation as forester. In 1928 he became superintendent at Stirling City and later was appointed production manager, a position that placed him in charge of all logging and lumber production at Stirling City and Chico. In 1946 he resigned this position to join the National Lumber Manufacturers Association in Washington, D. C. as their Executive Secretary. I am indebted to Mr. Colgan for reading the first draft of the manuscript and for a number of valuable suggestions on it. His help on describing the construction and operation of the huge double incline on the logging railroad during 1927 to 1935 was especially valuable. The double incline was Mr. Colgan's idea to solve a difficult logging engineering problem and he had a major hand in guiding a consulting engineer on its design.

The help given to me by Stanley Keesy, former logging railroad trainmaster and logging engineer, goes back to 1963 when I started to research an article on the logging railroad. This article, intended to follow an article on the Butte County Railroad in *The Western Railroader,* was never submitted due to its length.* Mr. Keesy has been an excellent source for information about the logging railroad from the mid-1920's through its abandonment, and the author recalls a number of very enjoyable evening interviews. I am also indebted to him for checking the location of railroad grades on a map loaned by the corporation, and for the identification of a number of photographs.

The logging railroad chapter on the standard gauge era and the car roster were enriched by the recollections of the late Harold L. Beavers, who also helped with the identification of numerous photographs. Mr. Beavers was engineer on the McGiffert log skidder, called the "jammer" by the employees, for more than a decade. Then in 1942 he served as engine messenger

*An article on the logging railroad by the late Stanley T. Borden was subsequently published in *The Western Railroader,* September-October, 1970.

for bringing Shay No. 106 down from Oregon after Diamond bought it. No. 106 became his pet engine; he spent the next decade on her right-hand seat box until the railroad quit. I feel fortunate to have made Mr. Beaver's acquaintance and enjoyed his reminiscences – not only about Diamond but other companies that he had worked for at various times – Swayne Lumber Co., Clover Valley Lumber Co. and Central California Traction.

A number of other former employees have been helpful. These include Dana Bailey; Norman Batham; Rollin Batham; Paul Blain; Bill Breedlove; the late Ira E. Brink; Mr. and Mrs. George A. Brown; the late Frank A. Compton; Cliff Gilbert; Milbourne Hedge, Sr.; Stanley Hedge; Cliff Henry; Norman "Dutch" Hubbard; Ralph Hupp; Durrell McEnespi; Elmer Marion; the late Harry Marion; the late Ed Mason; H. A. Otis; Ernie Robinson; George W. Scott, Sr.; William Shippen; the late M. B. "Brad" Smith; and Ernest Van Sant.

Family members of former employees I am indebted to include Alma Blasingame; Josephine Bloxham; Mrs. Frank Clough; Mrs. Justa Farley, the late Mrs. Fred Haskins; Vera Hedge (Mrs. Milbourne Hedge, Sr.); Isabel Millington; Beryl Orner; Mrs. Leta Robbie; Mrs. Eda Seaman Scott; Mrs. Edna Rouse Thelander; and Mrs. W. G Wahl.

I am indebted to several persons for help with the chapter on the Butte County Railroad – the late C. R. Dibble, who was one of the last surviving employees of the railroad, and Mrs. Lettie Lillybridge, daughter of J. F. Myers who was the fireman on that ill-fated freight train that piled up below Paradise in 1909. Former Southern Pacific employees who helped were Carl "Gil" Gilliland; Elmer Kennedy; Louis Ligon; the late Ernest Mullinix; and George W. Waite. I also wish to thank two employees of the Southern Pacific – Eldon Dyer for describing train operation on the Stirling City Branch following the Second World War, and George Kraus of the Public Relations Department.

Contributions from railfans are important for any railroad book and the list is long. I am very much in debt to David L. Braun for his excellent cartographic work on the maps and his interest in the project. For their contributions of photographs and/or information my thanks are due to Gerald M. Best; the late Stanley T. Borden; Ray Briggs; H. L. Broadbelt; John R. Cummings; Guy L. Dunscomb; Tom Gray; Bryan Griebenow; Francis A. Guido; Robert M. Hanft; Kirby M. Harris; Hank Johnston; Ken Kidder; John Labbe; Robert R. Lowry; Marvin Maynard; Warren Miller (Railway Negative Exchange); P. E. Percy; R. Stephen Polkinghorn; Douglas S. Richter; Al Rose; the late Lawrence W. Sagle; Vernon J. Sappers; Bill Shippen; Fred A. Stindt; Joseph Strapac; Harold K. Vollrath; Bert Ward; W. C. Whittaker and Ted Wurm. I owe a special thanks to Doug Richter and Joe Strapac for their help with the locomotive rosters.

Prior to working on the book itself, I wrote two articles on Diamond's railroads for railfan magazines. The editors of both magazines, Francis A. Guido of *The Western Railroader* and Karl R. Koenig of *Pacific News,* were kind enough to grant permission for a portion of the articles to be incorporated into the book if necessary. For this I am grateful.*

Many other people were helpful in various ways including Joseph F. Brouillette; Frederick "Bob" Stansbury Clough; Lloyd Cook; Sylvia Cook; the editors of *Forest Industries* (formerly *The Timberman*); Mildred Forester; Helen Gheller; D. W. Gutscher; Vivienne L. George; John Hamby; Dorothy Hill; Mary Lemcke; Leo A. McCoy; Mrs. Archie McDonald; Dr. Clarence McIntosh; John Nopel; Andrew Osborne; Henry J. Schaffer; Julia M. Shaffer; Paul M. Smith; the late Angeline Stansbury; Florence H. Styles; Ruby Swartzlow; and A. H. Turk. I am also grateful for the help of librarians and other staff members of several institutions; Mary S. Pratt of the Los Angeles Public Library; Don Kunitz of the Special Collections Department, University of California Library at Davis; Rose M. Daly of the California State Archives; Dr. Norris A. Bleyhl of the California State University, Chico Library; and the staff of the California Room of the California State Library.

I am very much in debt for the help given by W. H. Hutchinson, who read the first draft of the manuscript. Any errors, however, are my own.

Thanks are also due to fellow railfan, Cliff Baumer and his wife Sally, who knew of my special interest in Diamond and came back from a camping trip with some spikes from one of the abandoned logging railroad grades.

Last, but not least, a big thanks to my wife Joanne, who has lived with this project for a number of years, patiently read several drafts of the manuscript and aided with the correction of the proofs. She also helped by sending me back to my typewriter when I procrastinated.

KENT STEPHENS,
Chico, California.

*The articles were "Butte County R.R., S.P. Stirling City Branch" in the July and September, 1963 issues of *The Western Railroader* and "Diamond National Corporation" published in *Pacific News,* June 1964. The latter article focused on the plant switching railroads at Chico and Red Bluff.

CHAPTER ONE

Gold and Lumbering

THE FASCINATING story of the Diamond Match Company's California lumber operations and its colorful Butte County Railroad in one sense goes back to January 24, 1848. On that date, an event in the Sierra Nevada foothills changed the course of California – and even American – history. James Wilson Marshall, a master carpenter employed by John Sutter, discovered gold in the tailrace of the sawmill he was building at Coloma. The spread of the word of discovery and the subsequent mass movement westward of the "Forty-niners" is a well-known story. Few people realize, however, that the gold discovery came because of John Sutter's quest for lumber.

Sutter's sawmill at Coloma operated for less than a year, when it closed permanently. The gold rush, nevertheless, created a large demand for lumber and many sawmills sprang up throughout the Sierra Nevada gold fields.

Some miles to the north of Sutter's sawmill, gold was discovered July 4, 1848, at a large sandbar on the Feather River a few miles north of present-day Oroville. The site quickly became known as Bidwell Bar, honoring its discoverer, John Bidwell. Immigrating to California in 1841, Bidwell for the next six years worked for Sutter in various capacities – including ranch foreman and aide-de-camp during military campaigns. He visited the Feather River area while searching for a sawmill site for Sutter prior to the construction of the Coloma mill. The area to the north of the Feather River, within the Sacramento Valley, so impressed Bidwell that he purchased a Spanish land grant. This was William Dickey's 22,214 acre "Rancho del Arroyo Chico," on the north side of Chico Creek.

Two years after Bidwell's gold discovery, California entered the Union as the 31st state with its 27 original counties. Both Bidwell Bar and "Rancho del Arroyo Chico" were in Butte County, at the time much larger than now. John Bidwell was elected to the first State legislature, and hoped the county seat of Butte County would be near his ranch or in the southern portion of the county. The voters, however, selected Bidwell Bar where the county seat remained until 1856. Another election transferred it from the rapidly declining mining camp to the newer and growing town of Oroville, a few miles downstream on the Feather. Oroville has remained the county seat.

Bidwell founded the town of Chico in 1860 on a portion of his ranch, and offered free lots to anyone who would settle in his town. Within a decade Chico had a population of 2,000. Bidwell's home was on the edge of town. "Rancho del Arroyo Chico" became famous for its agricultural productivity and experimental nursery. Bidwell entered national politics in 1892, running for President of the United States on the Prohibition Party ticket, but polled only 271,058 votes. Bidwell passed away in 1900 and his wife died in 1918, but the name Bidwell is remembered in Chico. Before her death, Mrs. Bidwell gave 2,400 acres along Big Chico Creek to the city of Chico for a park. Only two municipal parks in the nation are larger.* The family home, Bidwell Mansion, is a state historical landmark.

The gold rush brought lumbering to Butte County, and the 1852 Census listed fourteen sawmills in the county. All but one was on the Feather River, close to the gold diggings. Eight years later the 1860 Census

*Largest municipal park in the nation is 4,253-acre Griffith Park in Los Angeles. Second in size apparently is Fairmount Park in Philadelphia (approximately 3,845 acres).

One of the earliest-known photos of Bidwell Mansion shortly after completion in 1868. The mansion contained 27 rooms and cost $60,000. (John Nopel collection)

The Bidwell Mansion at the turn of the century. It is now a state historical monument. (John Nopel collection)

General Bidwell gave land from his holdings for Chico Normal School, today California State University, Chico. This three-story brick structure was the main campus building until it burned in 1927. (James McK. Wright)

General John Bidwell, founder of Chico, (John Nopel collection)

showed one less sawmill, but production amounted to 14.5 million board feet. Prime evidence that lumbering was directly tied to gold mining was the fact that neighboring Tehama County to the north – out of the gold fields – had only two sawmills with an annual production of 1 million board feet.

All of this pine lumber was consumed locally. This was due partially to the transportation problem. Sawmill operators of this era faced major difficulties in moving lumber to the floor of the Sacramento Valley. Teaming costs were as high as $13.00 a thousand feet over rough mountain roads. In the summer, dust lay inches deep on the road, which the first rain turned to a sea of mud.

Even after the lumber reached the valley floor, the transportation difficulties were not over. River drives never attained much importance in the Northern Sierra Nevada, as the wild and unpredictable rivers were virtually impossible to drive either logs or lumber. The Sacramento River wasn't much better since it was extremely crooked and hence difficult for rafting lumber.

Much lumber was shipped on the river steamers plying the Sacramento River as far north as Red Bluff. In 1867 alone, the lumbermen shipped 799,927 feet of lumber from Chico Landing. Shipments of lumber on the Feather River, however, amounted to much less. This was because the Feather never attained the importance for river navigation as did the Sacramento. Even with the relatively cheap river transportation, lumbermen of the northern Sierra found it impossible to compete with the coastal sawmills in supplying the cities of Sacramento and San Francisco Bay.

Shortly after the close of the Civil War, both parts of the transportation problem were solved for the lumbermen. The California & Oregon Railroad, today a part of the Southern Pacific, was built north through the Sacramento Valley – arriving at Chico on July 2, 1870. The railroad not only provided economical and dependable transportation, but also consumed large quantities of cordwood.

The cheaper transportation from the mill to the railroad came through the use of V-flumes. The first V-flume was pioneered in Nevada by J. W. Haines, to move lumber to the Comstock Lode. It was pioneered in Northern California by C. F. Ellsworth.*

After operating three different mills over a four year period, Ellsworth purchased two neighboring Tehama County mills: the "Old" Champion Mill from Herbert Kraft in 1870, and the Belle Mill in 1871 from Kellum Powell. It was from the latter mill, situated in Belle Meadows near present day Lyman Springs, that Ellsworth started building his V-flume.

It took three years for the flume, locally called "Ellsworth's folly," to reach its valley terminus at the long-since vanished town of Sesma. The flume confounded critics by being almost an instant success. Ellsworth unfortunately did not live to see vindication of his "folly:" he died after falling from the flume while supervising its final construction.

The completion of Ellsworth's flume opened the V-flume era in Northern California. Even before it was finished there were two other flume projects – one each to the north and south of it. The flumes may have solved a transportation dilemma, but they were expensive and the economic stresses of the mid-1870's brought all three companies to the brink of financial

*A V-flume was a wooden water trough shaped like a V: they were used to transport lumber or logs to their destination. Most of the V-flumes in northern California were similar to Ellsworth's flume. Their approximate dimensions were around 48 inches in width at the water level narrowing to 16 inches at the bottom. The flumes were built in 16-foot long sections, called "boxes," which rested on cribbing or trestles depending upon the terrain.

insolvency. In 1876 their owners sold out to Comstock multi-millionaire Alivinza Hayward and his partner, former Great Lakes lumberman Norton Parker Chipman. They formed the Sierra Flume & Lumber Company.

Hayward and Chipman followed a policy of expansion that resulted in the Sierra Flume & Lumber Company becoming the largest single lumber operation in the world by 1877. The company used 23 miles of logging tramway to feed logs to 10 sawmills, and 156 miles of flume to transport lumber to the valley. Communications for operating this vast enterprise required building the largest private communications system in California: 250 miles of telegraph line linking mills, valley terminals, and offices. During the first year of operation, hardworking crews produced 51 million board feet of lumber; they used 8 million feet for company construction projects. For the first time, northern Sierra pine lumber became competitive in California cities with that from the coastal mills.

History again repeated itself. The company had expanded too fast in the unsettled conditions following the Panic of 1873. Another financial panic in San Francisco spelled disaster for Sierra, which went into bankruptcy during 1878. It was reorganized under the name of Sierra Lumber Company, which purchased the old company's assets. The same interests connected with its predecessor initially owned it, but they attempted to sell part or all of the firm during the next few years. Control of Sierra passed in the early 1880's to John C. and Edward Coleman of San Francisco. The English-born brothers made a fortune from gold mines in the Grass Valley region and helped found the Nevada County Narrow Gauge Railroad.

Far from California, an event on December 3, 1880 on the East Coast would have profound influence on Sierra Lumber Company 27 years in the future. On that day the owners of twelve match companies met in New Haven to consolidate their firms under the name of the Diamond Match Company of Connecticut. Two companies took the lead in forming the association for the common good of the match industry following a disastrous price war between them. These were Swift & Courtney & Beecher of Wilmington, Delaware, and the

The river boat Dover *on the Sacramento River. (James Morehead photo from Dorothy Hill collection)*

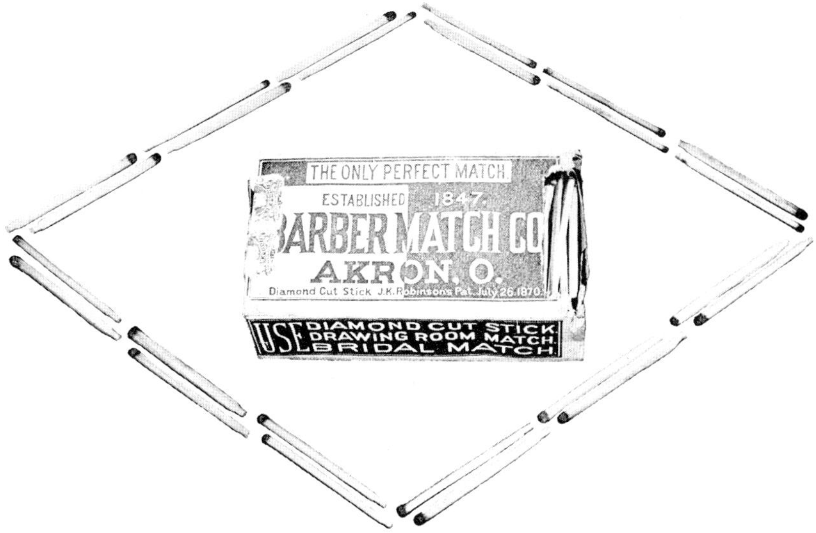

The diamond-shaped match splint was patented on July 26, 1870, by John K. Robinson (above left) of the Barber Match Company. When Diamond Match Co. was formed in 1881, it took its name from the diamond-shaped splint. (From Diamond Digest)

Barber Match Company of Akron, Ohio. The new company received its name from the diamond-shaped match splint (i.e., the match stick of the common kitchen match, called the S.A.W. match – "strike-anywhere match"), an important technological advance within the industry that had been invented by John K. Robinson of the Barber Match Company almost a decade earlier.

Diamond's first president was William Swift, who was the president of Swift & Courtney & Beecher at the time of the merger. He served until his retirement in 1889. His successor was Ohio Columbus Barber, son of the founder of the Barber Match Company.

Returning to C.. F. Ellsworth: he not only pioneered the V-flume in Northern California, but also built the first logging railroad – more properly called a logging tramway. While building the flume, his Belle Mill began to run short of a supply of logs within its cutting radius. Ellsworth realized he faced the choice of moving the mill and extending the flume to the new mill site or devising some means of transportation of logs to the mill. Thus resulted the building of the first logging tramway in the northern Sierra in 1872.

Ellsworth's crude tramway utilized peeled logs for rails laid on a ninety inch gauge. (U.S. rail gauges were being standardized then at the narrower four feet 8½ inches, and many logging railroads used a 3-foot gauge.) The logs were hauled on huge four wheel "trucks" pulled by oxen. The Marysville Foundry Company cast the iron wheels, which were forty-two inches in diameter and weighed 1,100 pounds each. A twelve-inch concave face kept the wheels on the log rails.

The logging tramway was no longer unique four years later when Sierra Flume & Lumber Company entered business. Most sawmills acquired by Sierra boasted of a tramway or got one through the new firm's initial improvement program. Once the logs were cut in the woods, they were yarded* to the tramway on large, four-wheeled "trucks" (i.e., wagons) pulled by oxen. At the tramway landing, the logs were loaded on the tramway "trucks" by the cross-haul loading technique. When several "trucks" were loaded, they were "roaded" to the sawmill behind oxen.

In the early months of 1881, logging contractor George Winans signed an agreement with Sierra to operate the "Old" Champion and Yellow Jacket mills at the head of Ellsworth's flume on what was then known as the Antelope Division. The agreement provided for rebuilding and extending a logging tramway for operating a light locomotive. Again it was a pioneering effort, as Winans and Sierra built what is belived to be the first logging railroad in the northern Sierra.

The railroad, it was decided, would be built to the same gauge as that of the old tram cars. This would allow use of these cars and the locomotive on the same track and also would permit utilization of two miles of existing wood tramway out of the "Old" Champion

*Some logging industry terminology should be defined at this point. The process of moving the logs from the cutting site to the loading site was, and still is, called "yarding." The loading site, in this case alongside the tramway, is known as the "landing." This could be at any convenient point that the loggers could find some reasonably level ground.

Ellsworth's "folly" terminated at the long-since vanished town of Sesma, on the east bank of the Sacramento River, across from present Tehama.

Ellsworth's Belle Mill in 1876. Ellsworth built the first V-flume in northern California in 1871-73. (Both Will Taylor drawings, Tehama County Library collection)

Mill. When the track was first measured, the gauge was determined to be 40½ inches. Afterward it was discovered this figure was a mistake as the tram cars were different gauges and the two miles of track varied from 39 to 41 inches. The specifications were then revised to the closest figure of the actual gauge, determined to be 39.35 inches, or European meter gauge. The apparent reason for the choice of this unusual gauge within the United States stemmed from a county ordinance that taxed railroads having a gauge commonly used in this country.

The locomotive, a small O-4-OT saddletanker, was built by Prescott, Scott & Company of San Francisco. It arrived in Red Bluff on a standard gauge flat car on April 30th; that was the beginning of difficulties between Sierra and Winans over the contract which ultimately ended in court. Upon inspecting the engine, Winans claimed it was incorrectly designed for operation on T-rail instead of strap iron as ordered. The manufacturer claimed the engine would be satisfactory and insisted the company should try it. The logging railroad experienced operating difficulty throughout the season, lowering Winans' lumber production. The locomotive was expected to operate on rails made out of three-by-six wood scantling faced with strap iron. It made a poor showing, continually breaking the strap iron under its weight with more or less constant derailments. The company finally ordered 20-pound iron rail in desperation, and ordered Winans to relay the tramway. At the end of the season, Winans was paid for the lumber he had produced. Subsequently he disputed the contract and sued Sierra. He won in Tehama County Superior Court, but the California Supreme Court reversed the judgment to favor Sierra when the firm appealed.

The logging railroad was rebuilt, adhering to the 39-35-inch gauge, and operated successfully. Beginning in 1883, Sierra purchased larger O-6-OT saddletankers from Porter, one by one, for the lengthening logging railroad. A second meter gauge logging railroad was built in 1899 out of the West Branch Mill in Chico Canyon on the Chico Division.

Sierra also operated two standard gauge railroad lines.. Both dated from the 1870's and connected a

Top: Ellsworth's flume in the "Hog's Back" gorge. Note the man riding the flume boat. This was the quickest way to the valley, especially in an emergency.

Bottom: The flume east of what today is known as Finley Lake. It can be easily seen why flume construction was so costly. (Both Will Taylor drawings, Tehama County Library collection)

Close-up of a typical homemade flume boat. The passengers were Dr. Newton T. Enloe, his sister Emma, and her family. This was the Chico Division flume at the edge of Chico. (Mrs. Justa Farley collection)

factory with the California & Oregon Railroad. The spur at Red Bluff traversed Oak Street from the railroad but shared the Centennial Free Bridge over the Sacramento River with a road. Each time the train crossed the bridge a flagman stopped the horses and wagons. Initially horses pulled single cars on the line, but a small locomotive was purchased in 1877 from the Marysville Foundry Company. This little engine, probably the only one ever built at Marysville, served faithfully until after the turn of the century when it was replaced by a new Climax geared locomotive. The company then moved it to Chico for use on the spur connecting the Chico division factory with the railroad.

The 1880's were prosperous years for Sierra and by 1891 the Red Bluff plant was the largest pine manufacturing factory in the world. The Colemans operated the company along conservative lines for an annual income rather than the headlong speed that characterized its predecessor. Much of the company's success was due to capable men, from President Peter Dean to the various superintendents and managers in the mills and yards. By the turn of the century, however, Sierra found itself beset with new pressures. Much of the timber accessible with the equipment of the 1880's was exhausted or in limited supply. Major timber purchases were being made on all sides, as large timber interests sought new sources of lumber. Most notable among these were the Curtis-Collins-Holbrook partnership and Thomas B. Walker, founder and president of the famed Red River Lumber Company. Additional large acreages were being withdrawn by the Federal government for national forests. These pressures, coupled with the death of Peter Dean in 1901 plus their own advanced years, resulted in the Colemans deciding to sell Sierra when the right opportunity arrived.

Above: Sierra's Chico Division flume shown (left) at the mouth of Big Chico Canyon, and (right) on a trestle along Centennial Avenue. (Both D. I. Corp. collection)

The original Sierra Flume & Lumber Co. plant in Chico (below). The flume was pulled by hand. (Will Taylor drawing, Tehama County Library collection)

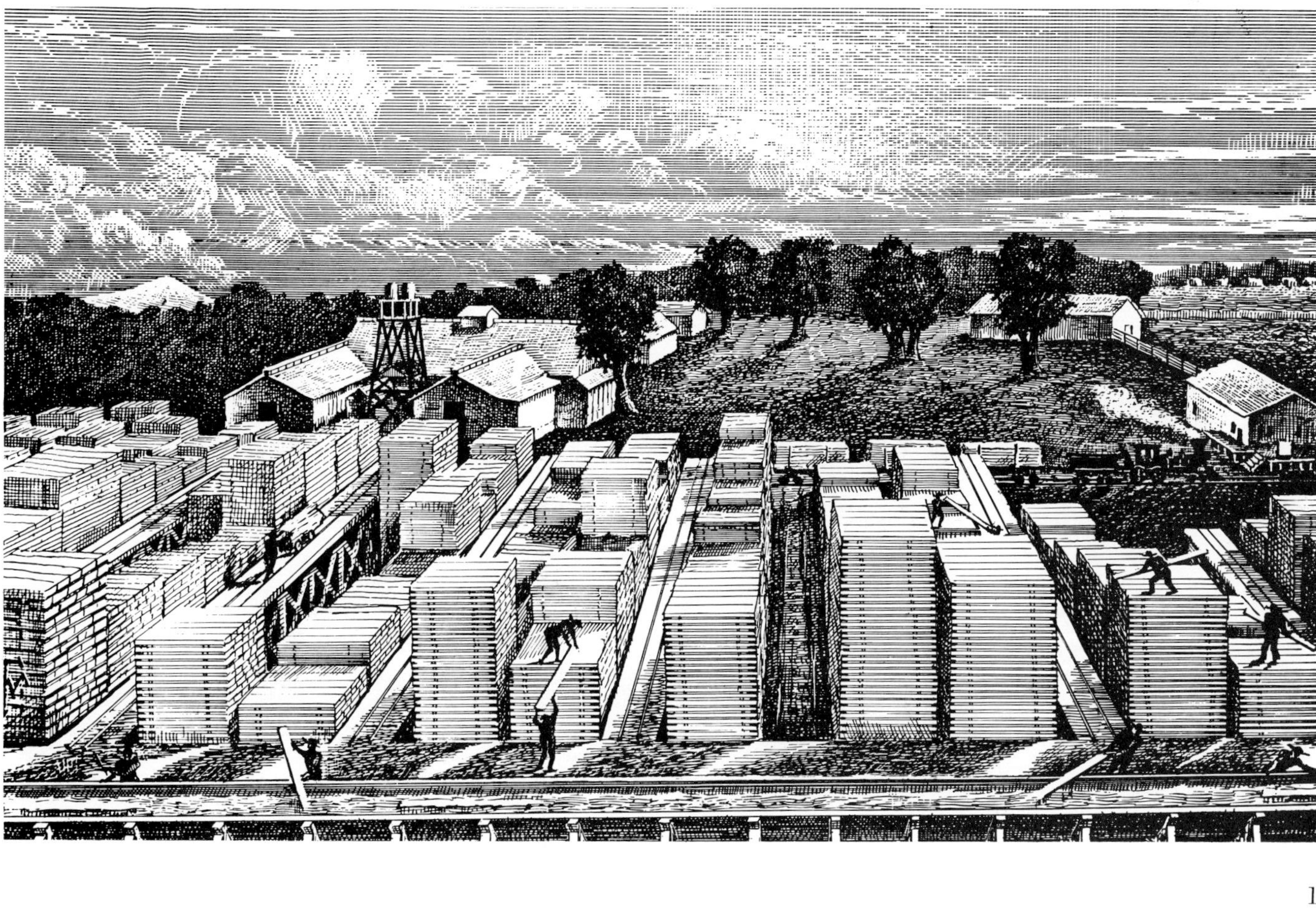

Sierra's Red Bluff plant and the Centennial Free Bridge in 1876. The flume into the plant can be seen on the left side of this artist's drawing.

The Moscow Mill drawing below shows logs being hauled by horse teams and wagons to the sawmill. (Both Will Taylor drawings, Tehama County Library collection)

Roading logs to the Belmont Mill in 1876. A brakeman rides the second truck to control the speed. (Will Taylor drawing, Tehama County Library collection)

Sierra No. 3, the "Uncle Sam," brings a six-car train to the "Old" Champion Mill. Note the size of the big sticks on the cars. (D. I. Corp. collection)

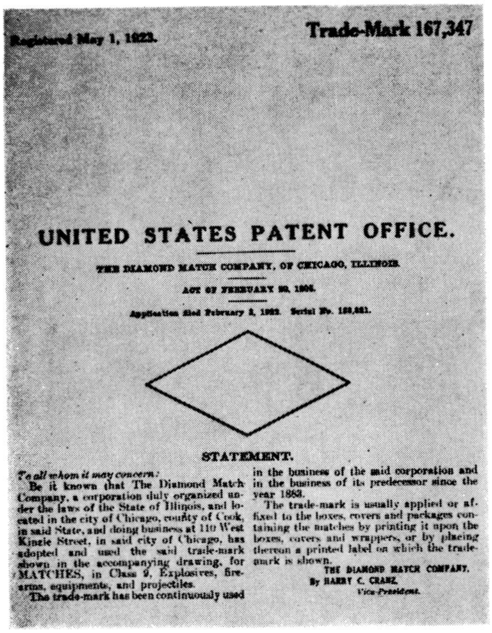

Registered May 1, 1923. Trade-Mark 167,347

UNITED STATES PATENT OFFICE.

THE DIAMOND MATCH COMPANY, OF CHICAGO, ILLINOIS.

ACT OF FEBRUARY 20, 1905.

Application filed February 3, 1922. Serial No. [illegible].

STATEMENT.

To all whom it may concern:

Be it known that The Diamond Match Company, a corporation duly organized under the laws of the State of Illinois, and located in the city of Chicago, county of Cook, in said State, and doing business at 110 West Kinzie Street, in said city of Chicago, has adopted and used the said trade-mark shown in the accompanying drawing, for MATCHES, in Class 9, Explosives, firearms, equipments, and projectiles.

The trade-mark has been continuously used in the business of the said corporation and in the business of its predecessor since the year 1883.

The trade-mark is usually applied or affixed to the boxes, covers and packages containing the matches by printing it upon the boxes, covers and wrappers, or by placing thereon a printed label on which the trade-mark is shown.

THE DIAMOND MATCH COMPANY,
By HARRY C. CRANE,
Vice-President.

1883-1911

1911-1921

1921-1953

1921-1953

1953-1956

1956-1958

1958-1959

1959

1960-

A diamond-shaped symbol, or logotype, has been used by the Diamond Match Co. (now Diamond International Corporation) as part of its corporate identity since 1883. These illustrations show the changes over nearly a century.

CHAPTER TWO

Newcomer to the Sierra

The Diamond Match Company

ON JUNE 19, 1901, one J. H. Comstock and his wife arrived in Chico, California, to spend several weeks while he cruised* the almost 40,000 acres of standing timber in Butte, Plumas, and Tehama counties that comprised the Sierra Estate. Comstock's employer was The Diamond Match Company and he was head of that company's Michigan timber operations.

The history of the Sierra Estate goes back to 1889, when Frederick H. Deakin (now spelled Daken) signed a contract with the Central Pacific Railway to purchase over 50,000 acres of the railway's timberlands near Chico. These timberlands had been granted to the railroad from the public domain as construction subsidies. Deakin, apparently unable to complete the purchase on his own, took in Thomas A. C. Dorland of Alameda as a partner. Even with a partner the terms were too expensive and the contract was greatly reduced in 1892. In 1897 Dorland passed away and his estate's interest in the contract went to the Central Pacific. The same year the Sierra Estate Company was incorporated "to acquire, hold, deal, etc., in mining properties, lumber properties and real estate." The five incorporators included Frederick H. Deakin.

Additional purchases of timberlands brought the total acreage close to 40,000 acres by 1900. Deakin then made the first effort to sell the property to a lumber company, reportedly offering it to timber baron Frederick Weyerhaeuser, Sr. The Weyerhaeusers dispatched a cruiser to inspect the timber. The quality of the timber impressed the cruiser, but he purportedly wanted a $10,000 personal fee for making a favorable recommendation. Deakin refused and the cruiser turned in an unfavorable report. Weyerhaeuser declined to purchase the property and in 1901 it was offered to The Diamond Match Company.

The Sierra Estate was one of two tracts of timber being considered at the time by Diamond's Board of Directors. The other tract, in Canada, was cruised by Comstock during the winter of 1900-1901. He never made a report on the Sierra Estate, however, for he was fatally stricken with cerebral meningitis while at Chico; he died at his home in Green Bay on August 6, 1901. W. M. Graves, then general manager of Diamond, reportedly went to Green Bay to see Mrs. Comstock, who told him of her husband's enthusiasm for the property. Whether this was responsible for increasing Diamond's interest in the Sierra Estate is not known, but on September 28 Frederick H. Deakin, Fred M. Clough, John K. Robinson and Edward T. Hall arrived in Chico to stay at the Park Hotel. Clough and Robinson were both connected with Diamond: Clough at this time held the position of traveling machine inspector plus being a general trouble shooter. Robinson, who was a veteran Diamond Match official, was treasurer of the corporation. The connection of Hall with the group is unknown, as he was not an official of Diamond. He may have been a timber cruiser hired by Diamond on a short term basis.

Deakin by then had an option on the Sierra Lumber Company's Chico Division timberlands and plant, and he attempted to sell it to Diamond at the same time as the Sierra Estate. He failed to interest Diamond in the Chico property but was successful in selling the

*Cruise is lumber industry terminology for taking an inventory of the value and quantity of timber within an area. The cruiser is the estimator of the inventory.

The Barberton, Ohio, plant of the Diamond Match Company shortly after the turn of the century. This was one of the largest match factories in the United States, producing 200 million matches daily. (Twenty Seven Years)

Sierra Estate. Diamond paid Deakin and his partners $318,992 for their 39,874.60 acres of timberland. Purchases of timberlands from other sources brought the total to 69,148.07 acres. The additional acquisitions tended to block out the checkerboard locations made by the Sierra Estate Company.*

Purchase of these timberlands gave Diamond an estimated standing 2 billion board feet of sugar pine, white (ponderosa) pine, fir, and other species of timber in this tract alone for an estimated yield of 30,000 board feet per acre. Several explanations have been offered as to why Diamond decided to purchase California timber.

One legend is that purchase of the timberlands was to obtain a source of match plank to replace the nearly depleted forests in New England and the Lake States that supplied Diamond's factories. This legend holds that lumber from the California tract was belatedly discovered to be unsatisfactory for match plank, forcing Diamond into the general lumber business and a search elsewhere for match plank. This legend, however, is refuted by evidence that President O. C. Barber and other top officials knowingly purchased a mixed-conifer forest containing a large percentage of timber other than sugar pine, the only suitable pine for S.A.W. ("Strike Any Where") matches. A statement on the purchase in the 1902 annual report mentioned only the transportation and distribution of lumber cut on the tract. In addition, plans for the sawmill two years later called for an annual cut of between 60 and 70 million board feet. The match business could not consume such a cut nor did it warrant the construction of the contemplated plant. Diamond, already in the lumber business when it bought the California tract, for years had operated a lumber department in Illinois and Wisconsin.

An almost forgotten fact is that Diamond did not invest in California solely for its own benefit. Diamond's 50-50 partner was Bryant & May Ltd. of London. This firm, incorporated in London on June 12, 1884, had a large portion of the sales until Diamond's entry into the English match market in 1894. Diamond's British match factory reportedly was the largest in the world and the competition drove Bryant & May's annual twenty per cent dividends down so low that the English company accepted a proposal from O. C. Barber that Diamond take control of the firm and guarantee an annual fourteen per cent dividend.

Bryant & May merged in 1901 with The Diamond Match Company of Liverpool Ltd. This merger gave Diamond a large block of voting stock, if not actual control of the company. Bryant & May's involvement in California with Diamond might have resulted from Diamond's control of the English match market, or it may have been that Diamond's large share of voting stock insured joint participation.

*Checkerboard locating was the favorite method of early day promoters to gain effective control over an area nearly double their own holdings.

Looking east on Third Street about 1900. Both the Odd Fellows Hall (three-story building with porch) and the water tower are still standing. (John Nopel collection)

The northeast corner of Fourth and Main was occupied for years by the Park Hotel. One of Chico's largest hotels, it was built prior to 1900. (John Nopel collection)

A stage and passengers pause for a photographer on the Humboldt Road about 1900. This road was built between Chico and Idaho in the 1860's. (John Nopel collection)

A turn-of-the-century view of Broadway in Chico's business district. Building on the corner housed the Bank of Chico. (John Nopel collection)

Diamond was vested with managing the California property in the best interests of both companies. Diamond or its employees held titles to purchases for several years. This became increasingly a point of contention between the partners, and in 1905 Bryant & May received title to a half interest in the property.

Active work on the development of the California property was under way early in 1902 under the direction of Fred M. Clough, who became Diamond's first Pacific Coast manager and greatly influenced the direction of the California operations during the initial years.

Mr. Clough was born at West Mount Vernon, Maine, on December 30, 1853. He began his working career as a machinist in 1873 and in the next thirty years had varying experience as a machinist, foreman, match manufacturer, and lumberman. He joined Diamond in 1893 as manager of the St. Louis factory: this was the first of increasingly responsible management positions he held with the company.

With his appointment as Pacific Coast manager, Mr. Clough drew upon his background and varied experience to establish a major manufacturing operation. Indeed, within the Clough family he was known as a "starter" who moved on once something was established; his family had moved on the average once every six years.

Mr. Clough established headquarters at the Union Hotel in Oroville for preliminary work on the California property. Everyone connected with the preliminary work initially expected Oroville to become the permanent headquarters for Diamond's California operations. This location could have been selected just as easily as several other valley sites. The flume method of transporting lumber was under consideration and a flume could go downhill to the valley by several different routes. Clough, however, was not impressed with this established method of lumber transportation in the Sierra. The idea of building a flume was soon discarded in favor of constructing a railroad into the mountains. Depending upon the sawmill's location, the railroad could be built from existing rail connections at Oroville, Durham, or Chico.

A site for the sawmill was finally selected at a point on Magalia Ridge, twelve miles above Paradise. The mill was located within the Sierra Estate timberlands at an elevation of 3,525 feet. A small draw provided a natural log pond without much construction expense: this may have influenced Clough's choice of the site.

Once the site for the sawmill had been determined, it was apparently left to Mr. Clough to decide where the manufacturing plant would be built in the valley. Chico won the industrial prize away from the neighboring towns by a community subscription to purchase the necessary land, then offered to the company at a lower price than competing sites. Purchase of the land began in October 1902, and on November 1 the deeds passed to Clough. The initial purchase totaled 241 acres just south of Chico next to the Southern Pacific (California & Oregon) tracks. The total cost to the Diamond-Bryant & May partnership was approximately $100 per acre, with thirteen acres set aside exclusively for Diamond's use as a match factory. (The match factory was the only part of the initial venture that was not jointly owned.) Rather than extending the name of Chico to the new plant site, it was named Barber in honor of O. C. Barber.

O. C. Barber was president of Diamond when the company purchased the Sierra Estate. (Twenty Seven Years)

Fred M. Clough, Diamond's first Pacific Coast Manager. (Frederick "Bob" Stansbury Clough collection)

Another major match factory owned by Diamond was at Oshkosh, Wisconsin. Other factories were situated at Portland, Maine; Detroit, Michigan; and Oswego, New York. (Twenty Seven Years)

Diamond's timberland purchases included some large trees. (Twenty Seven Years)

Early-day crew tracing section lines in Diamond's timberlands. (Twenty Seven Years)

CHAPTER THREE

Development of the California Property

DIAMOND'S first major construction project was building a railroad between Barber and the sawmill site on Magalia Ridge. Completion of a railroad between the two points would enable Diamond to use its own lumber for much of the construction work at Barber, and the lack of a railroad would conversely make it difficult to move the heavy sawmill machinery to the mill site.

On September 18, 1902, Clough was authorized by Diamond's Board of Directors to obtain a franchise for a firm to be known as the Butte County Railroad Company. Acting as an individual, Clough went before the Butte County Board of Supervisors to request a franchise for the railroad. The Board granted a franchise on November 11 to allow the Butte County Railroad to construct its line into the mountains from a point on the Southern Pacific between Chico and Durham. A second franchise obtained on January 8, 1903, amended the first one by authorizing the railroad to build from a point on the Sacramento River. Clough and other officials apparently envisioned that they might have trouble with the Southern Pacific. Building from the river would allow the lumber to be shipped via steamer should rail problems arise.

On February 4, 1903, the contract to build the railroad was awarded to E. L. and A. B. Stone of Oakland. The contract called for completion of the railroad to the sawmill site by September 1, 1903.

The official incorporation of the railroad occurred a few days later. The Butte County Railroad Company was incorporated under the laws of New Jersey on February 23, 1903, with offices at 83 Montgomery Street in Jersey City. Capitalization was set at $1 million, divided into 10,000 shares at the par value of $100 each. The largest stockholder of record was Russell Hawkins, assistant to President Barber of Diamond, who held 410 shares of the 418 issued.

Clough in early 1903 hired John B. Robinson from the Southern Pacific to serve as superintending engineer and general manager of the railroad. He was joined by another former S.P. official, Frank G. Sumner, who resigned as road master to join the Butte County as assistant engineer.

The contractor moved the first dirt in grading the roadbed at Paradise on March 5 and established a second camp at Butte Creek the same day. By March 16 three camps had been established along the line and the contractor was waiting for the weather to clear before activating three additional camps.

The first five miles of the roadbed between Chico and the foothills was easy going for the construction crews. The free soil was scraped by patent grader, operated by steam power and moved by a traction engine as the grade was completed. Once in the foothills, the contractor used scrapers to move the loose soil and rock carts to haul the rock. A steam shovel excavated the cuts.

Track laying began at Barber in late June; within a few days two miles of the track had been spiked down. Sufficient track materials, however, were available for only three miles of rails. Various circumstances delayed construction. Since no firm in the United States could fill the order quickly enough, the rails were ordered almost a year earlier from a firm in Antwerp,

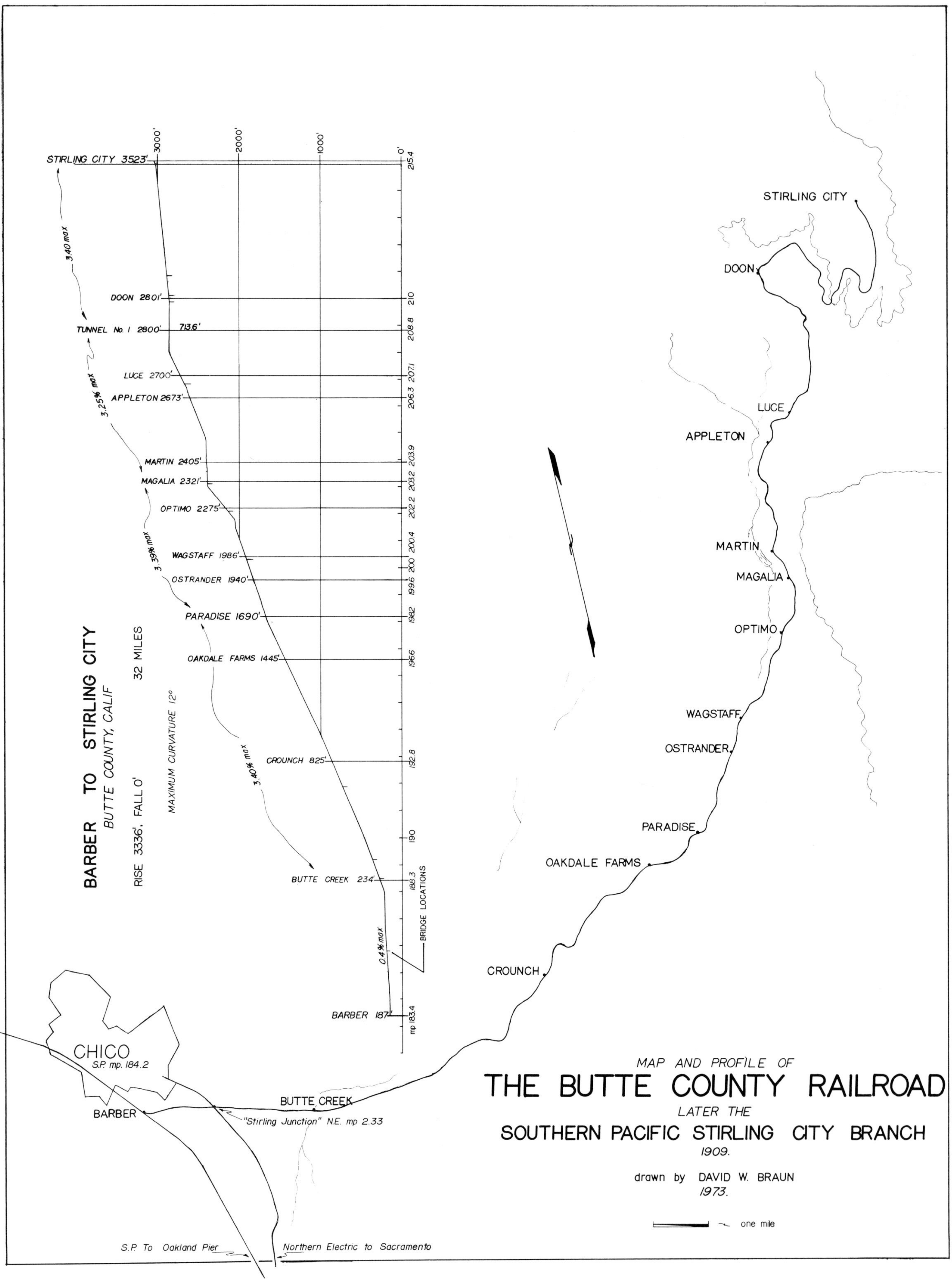
MAP AND PROFILE OF
THE BUTTE COUNTY RAILROAD
LATER THE
SOUTHERN PACIFIC STIRLING CITY BRANCH
1909.
drawn by DAVID W. BRAUN
1973.
one mile
BARBER TO STIRLING CITY
BUTTE COUNTY, CALIF
RISE 3336', FALL 0'
32 MILES
MAXIMUM CURVATURE 12°
3000'
2000'
1000'
0'
STIRLING CITY 3523'
215.4
3.40 max
DOON 2801'
210
TUNNEL No. 1 2800'
713.6'
208.8
LUCE 2700'
207.1
3.25% max
APPLETON 2673'
206.3
MARTIN 2405'
203.9
MAGALIA 2321'
203.2
OPTIMO 2275'
202.2
3.39% max
WAGSTAFF 1986'
200.4
OSTRANDER 1940'
200
199.6
PARADISE 1690'
198.2
OAKDALE FARMS 1445'
196.6
3.40% max
CROUNCH 825'
192.8
190
BUTTE CREEK 234'
188.3
BRIDGE LOCATIONS
0.4% max
BARBER 187'
mp 183.4
STIRLING CITY
DOON
LUCE
APPLETON
MARTIN
MAGALIA
OPTIMO
WAGSTAFF
OSTRANDER
PARADISE
OAKDALE FARMS
CROUNCH
CHICO
S.P. mp. 184.2
BARBER
BUTTE CREEK
"Stirling Junction" N.E. mp 2.33
S.P. To Oakland Pier
Northern Electric to Sacramento

Butte County R. R. Company

TRAIN TIME TABLE

Beginning Nov. 1, 1903

LEAVE	STATION	ARRIVE
7:15 a. m.	Barber	6:00 p. m.
7:27 a. m.	Butte Creek	5:48 p. m.
7:46 a. m.	Crouch	5:28 p. m
8:08 a. m.	Paradise	5:07 p m.
8:30 a. m.	Magalia	4:45 p. m.

Chico & Barber Bus.

Will leave Park, Union and Johnson House Hotels every morning at 6:45 to connect with train at Barber. Bus will connect with evening train also. L. SHERMAN, Prop

Butte County Railroad's first passenger train, the excursion to Magalia on Oct. 31, 1903, ready to leave Barber behind No. 2. (Pine Chips) *At left, the railroad's first timetable shows service only to Magalia.* (Chico Daily Record)

Belgium. In late June, they were on three ships that left Antwerp nearly 180 days earier for what presumably would be a 135-day voyage. Not until August 2 did the first ship, the German vessel *Butteshire,* arrive in San Francisco with a cargo of 1,450 tons of rails. The remaining 13,000 tons arrived even later.

The *Butteshire's* arrival enabled the railroad's officials to project that track would be completed as far as Magalia by September 15. But even if that date could be met, the railroad would be a little over half way to its terminus. Even this target date, however, was missed and it was mid-October before the track layers and the construction train reached Paradise.

A special excursion train carrying 300 company guests on a Saturday outing to Magalia celebrated completion of the railroad there. The train consisted of a coach and two flat cars, the latter being converted to open-air cars with the addition of railings and benches. It was powered by 4-4-0 No. 2, with Billie Orr at the throttle and A. J. Willis firing. At 1:00 p..m., as scheduled, Conducter Beamer shouted "All aboard!" and the train left Barber. A brief stop at the Butte Creek bridge showed the excursionists the construction work at that point. The train then continued on to Paradise, where it made a water stop before proceeding to Magalia. It arrived there by chance just as the track laying crew was driving the last spikes at the future site of the Magalia depot. That did not escape notice as the train moved onto the last rails and the track workers were cheered by the excursionists, marking the completion of 18.1 miles of the railroad and the operation of the first passenger train. The excursionists reboarded the train at 4:00 p.m. and they were back in Chico an hour later.

Regular train service between Barber and Magalia began the following Monday (November 2) with a scheduled roundtrip daily. At Magalia, the passengers boarded Perry's stage to ride the remaining miles to the sawmill site over rough mountain roads. The first scheduled passenger train carried neither mail nor express.

The railroad originally was scheduled to be completed on September 1 to the sawmill site. This schedule possibly could have been met had it not been for construction delays caused by an unexpected amount of lava encountered during grading. Another factor was the previously mentioned slowness of rail shipments. Planners also possibly did not foresee that a crooked dyke of lava near Lovelock would require a tunnel. Even not considering these costly delays, Diamond was spending huge sums on constructing the railroad.

The first agreement between the Butte County Railroad and the Southern Pacific was reached on May 23, 1903. While a traffic agreement, it laid the groundwork

A deep cut on the upper end of the railroad. It was no wonder the railroad wasn't completed in the alloted seven months. (Mrs. Eda Seaman Scott collection)

for negotiations for the eventual purchase of the railroad by the Southern Pacific. These negotiations had been under way for sometime when formation of the Chico & Northern Railroad was announced in the *San Francisco Call* on November 11.

Formation of the Chico & Northern caused excitement in the San Francisco and Chico newspapers, since many believed that the powerful Gould financial empire was backing the Butte County Railroad. This popular belief developed naturally. It was well known at the time that Diamond purchased the Continental Match Company from Edwin Gould, Jay Gould's son. As part of the deal, Diamond paid Edwin Gould 10,000 shares of stock and gave him a seat on the board of directors.

Thus the popular conception also developed that the Butte County Railroad served as a front for the aspirations of George Gould, Edwin's brother, to extend his rail empire to the Pacific Coast. At the time George Gould's ambitions were frustrated by E. H. Harriman, who then controlled the Southern Pacific, Union Pacific, and Central Pacific and had blocked Gould's westward expansion at Ogden, Utah.

Coupled to this conception of Gould influence in the Butte County Railroad was an extreme dislike of the Southern Pacific and Central Pacific railroads, whose dominance in California politics and transportation at the time was the subject of Frank Norris' book, *The Octopus*. In short, the Butte County Railroad represented the beginning of a deliverance from the cutthroat domination of the Central Pacific and Southern Pacific. The announced purpose of hauling logs and lumber was regarded as a blind to fool Harriman and hide the presumed beginning of Gould's transcontinental railroad.

The reaction of the newspapers to the incorporation of the Chico & Northern was predictable. It was cited as being the tried and true tactic of the Southern Pacific: build a parallel and competing line to freeze out the competitor!

But the big railroad fight was only in the minds of the newspaper editors. The final agreement to sell the Butte County Railroad to the Central Pacific was reached on October 27, 1903, prior to the formation of the Chico & Northern. The agreement provided for the C.P.'s immediate purchase of the partially completed railroad. The Butte County Railroad agreed to finish the line to the sawmill's site, and Diamond in turn would build a sawmill of not less than 150,000 board feet capacity per ten hour day. Satisfactory traffic agreements were worked out and the Butte County agreed not to build to the Sacramento River. The purchase price was set at the cost of the completed railroad — ultimately more than $1.5 million! Central Pacific agreed to lease the railroad line and equipment back to the Butte County Railroad for operation at an annual rental of five per cent of the total purchase price.

There were more banner headlines in the newspapers in late December when it became known that the Butte County Railroad had been transferred to the Central Pacific. One San Francisco newspaper carried the story under a headline reading "SURRENDERS RAILWAY TO HARRIMAN."

Construction work proceeded on other aspects of Diamond's California development simultaneously with the building of the Butte County Railroad. The first construction crew arrived approximately April 20, 1903, at the sawmill site and pitched tents among the pines. Almost coinciding with this arrival was the naming of the sawmill town, announced as "Sterling" in the April 22 edition of the *Chico Daily Record*. Two days later, the same newspaper described activities at the new town of "Sterling," but this time the spelling was "Stirling." No explanation was offered for the changed spelling and the town was thereafter referred to as "Stirling" or "Stirling City." It can be assumed that the first spelling with an "e" was an error — but one frequently made ever since.

According to legend, Fred Clough named the town "Stirling City" for The Stirling Company, an Ohio

According to legend, Fred M. Clough named Stirling City after the Stirling boilers installed in the sawmill's powerhouse. (Both photos, Joseph Brouillette)

based boiler manufacturing firm controlled by O. C. Barber. This company furnished the boilers for both the sawmill powerhouse at Stirling City and the one at the Barber plant.

Mr. Clough hoped to make Stirling City more than just a sawmill town and this hope was expressed in the pages of the *Chico Daily Record:*

> "Stirling City has often been mentioned as an ideal spot for a summer resort. Environed as it is with all that makes the hearts of summer pleasure-seekers glad; beautiful scenery, sparkling trout streams and convenient deer haunts; then its easy access when the Butte County Railroad shall have been completed . . . and the mild invigorating climate, should . . . attract hundreds of people from the heat and dullness of the valley . . ."[1]

Mr. Clough's plans for a combination summer resort-sawmill town probably influenced the location of the sawmill site and found support among Diamond's higher officials. Many of these executives joined Clough in private ventures connected with the development of the town.

The common belief in Butte County when Stirling City was founded was that it would be a typical company town, with Diamond monopolizing all commercial ventures. Mr. Clough denied this on several occasions, further stating that Diamond did not own or operate "company stores." The company would erect business buildings and offer them for sale to all "reputable businesses" to be operated by individuals.

This was true technically at Stirling City. Diamond did not own the major commercial ventures, but Diamond's officials did! These officials acted as individuals to form an independent corporation for each business. The same executives that controlled Diamond's local operations also controlled the independent corporations. As the businessses grew and proved themselves, an interest was usually purchased by several of the officials and directors in New York. Thus the company officers, rather than Diamond's stockholders, profited from the separate corporations. The economic terminology for such a firm is an "inside" corporation. Diamond was not the only company involved with inside corporations during this era, as it was a widespread practice in American business.

The most prominent inside corporation at Stirling City undoubtedly was the Stirling Mercantile Company. It was incorporated on April 17, 1903, with a capital stock of $35,000 divided into 350 shares. The five incorporators were John B. Robinson, W. P. Lynch, Charles A. Driess, J. F. Lynch, and J. F. Nash. A few months later there were additional stockholders, including Fred M. Clough, Major A. F. Jones, Florance Armbruster, E. D. Newcomb, and J. R. Graves. In

Clearing Stirling City townsite in 1903. Solid wheel log wagons were used to transport logs to the "Old" Bennett sawmill. (Both photos, Vera Hedge collection)

1906 O. C. Barber and Russell Hawkins joined the list of stockholders. All were connected with Diamond.*

The Stirling Mercantile Company initially operated out of several temporary tents in Stirling City. On June 15, 1904, the firm moved into permanent quarters in its own building. This was a three-story edifice on the northeast corner of Laurel and Quartz streets; it was the largest structure other than the sawmill. The first floor housed the mercantile store and Diamond's offices. The location of the latter in the rear portion of the building undoubtedly resulted in some business for the store. The company stocked groceries, hardware, mining supplies, blasting powder, coal, and all types of general merchandise. Drugs could be purchased at the store and a ready-to-wear and fine goods department was established. The company maintained a wagon delivery route to the surrounding area, including the logging camps.

*John B. Robinson (as differentiated from John K. Robinson, who was Diamond's treasurer) joined Diamond from the Southern Pacific: his position with Diamond was Chief Engineer, Butte County Railroad and logging railroads. W. P. Lynch located timber and worked in water engineering for the company. Dreiss was the son-in-law of the late J. H. Comstock and held the position of general manager of the merchantile company. J. F. Nash was the first superintendent of woods and sawmill operations at Stirling City: he had joined Diamond from the West Side Lumber Company. J. F. Lynch's connection with Diamond is unknown: he may have been involved with the timberlands. Clough was Pacific Coast Manager and the highest ranking official at Chico. Major Jones was a prominent Oroville attorney and local politican: he was retained by Diamond as their Pacific Slope attorney during the initial years at Chico and Stirling City. Florance Armbruster was the first auditor and head of the accounting department at Chico: previously he had been assistant secretary and auditor with Diamond at the New York headquarters. Ed Newcomb was manager of the mercantile company's store. J. R. Graves probably was Jack Graves, son of general manager W. M. Graves. Jack Graves worked for Diamond at Chico. Barber was president of the company and Hawkins was his assistant.

A number of Diamond's officials were also involved in a second important "inside" corporation at Stirling City. This was the Stirling City Bank, incorporated on August 21, 1903. The bank initially was in Chico, but moved to Stirling City once the town had been established. It was capitalized for $25,000 divided into 500 shares of $50 each. Its first directors were Fred W. Clough, John B. Robinson, Major A. F. Jones, W. P. Lynch and E. J. Lewis.* A number of other officials subscribed to its stock. Among them was O. C. Barber, who owned stock in both the bank and the mercantile company.

The bank's main function for Diamond was to handle its payroll as the company's paymaster. Once full operations were under way, the monthly payroll was substantial.

During the early months of 1903, the townsite was surveyed and a plat showing 332 lots was accepted by the Butte County Board of Supervisors. The lots ranged in price from $75 to $250 each and about $45,000 worth were sold during 1903.

A crew, supervised by Fred Haskins, started constructing the "big" mill at Stirling City almost as soon

*E. J. Lewis was the first head of Diamond's accounting department in Chico, serving until Florance Armbruster arrived from New York: later he was office manager of the match factory. He had formerly been Diamond's West Coast agent.

as the townsite was cleared of brush. Diamond's first California sawmill operations, however, were not at Stirling City but several miles below, at the Doon Mill. This facility was built by Mike Doon in 1869 and passed through several changes in ownership and location before it was sold to Diamond. It started sawing for Diamond in October 1902 under contract to Captain John Morrison, an old-time Butte County sawmill operator. The Doon Mill was a top-and-bottom circular mill; Morrison logged with a Dolbeer donkey and log chutes.

Diamond purchased a second sawmill in 1903. This was the "Old" Bennett Mill on Secret Creek, which was moved to a site just below the present log pond at Stirling City. The "Old" Bennett Mill started cutting in September 1903, with F. A. Bennett as its sawyer. The Doon Mill apparently was closed when the "Old" Bennett Mill started. The total cut of the Doon Mill was approximately 1.1 million board feet of lumber, of which nearly half went into the construction of the Butte County Railroad.

Preparing for winter months, crews in October constructed board sidewalks throughout the main part of Stirling City. Fred Haskins' crew then was well along with the basic building for the sawmill. Haskins had a driveway built in late October to the second floor of the mill, enabling the teams to deliver the construction materials upstairs without hoisting everything from the ground. The basic building for the sawmill was apparently finished by the end of 1903, but completion of the mill and the installation of its machinery was held up pending the railroad's arrival. This was possibly due to the high expense of teaming the sawmill machinery into Stirling City, but there were a number of available teamsters who could have moved most of the machinery to the mill had it been desired to rush it to completion.

Interior of Stirling Mercantile Co. store about 1907. Extreme left was Jack Lynch; third from left, George W. Jones; fourth from left, Walter Grace. (Alma Blasingame collection)

Construction work on the Butte County Railroad continued throughout the winter months. Grading crews made excellent progress during December's twenty-one clear days, but poor weather conditions slowed them in January. The grading was virtually completed by mid-February and only a little work remained on the tunnel. Bad weather in late February and early March continually hampered construction work. Despite the weather, the railroad reached the Doon Mill by mid-March; a few days later the residents of Stirling City could hear the whistle of the locomotive pulling the construction train.

The final push to reach Stirling City was not without difficulty, which dashed hopes of local residents to see the first train on April 1. The roadbed was a sea of mud during late March and early April, preventing

Stirling Mercantile Co. building when it was new. The first floor contained the store plus Diamond's offices. Second floor was shared by the mercantile offices and a Masonic Hall. Third floor contained sleeping rooms rented to company employees. (Florence Styles collection)

The Stirling City sawmill in 1904, shortly after it began sawing lumber. (Milbourne Hedge, Sr., collection)

the use of horses. Construction continued, however, due to the need to move the machinery for the sawmill from Barber where it had been stored all winter. Men worked in mud up to their waists laying the ties and rails on a foundation of planks and logs that was hoped would keep the new track from sinking completely out of sight. The track was laid into Stirling City on April 8, marking the town's railroad day. Despite the condition of the track between Doon Mill and Stirling City, the railroad started operating a daily train to move in carloads of machinery, groceries, furniture, and other merchandise — as well as to carry passengers. Cars jammed the small freight yard and the whole town bustled with activity as teams hauled supplies throughout the town. As soon as Stirling City was reached and a few rails laid for the freight yard, tracklaying started for the mill spurs.

By the time the railroad arrived in Stirling City, the sawmill was nearly ready for the installation of all of the machinery. Bricks for the boiler room were manufactured locally from a deposit of clay in Long Ravine near Stirling City. The ravine had a supply of wood to fire the kilns; Diamond apparently even considered going into the brick manufacturing business. Remains of the clay pit and discarded bricks can still be found in the brush in the ravine. Holt and Gregg, a Red Bluff contractor, erected the powerhouse and boiler room.

During July the sawmill machinery, manufactured by Allis-Chalmers, was installed in the mill with a representative of the company in residence to supervise the work. The powerhouse was complete enough in July to allow the installation of the Stirling boilers and the erection of four 120-foot high smokestacks. At the same time a 1,200 horsepower Corliss engine (rated 800 horsepower by its builder) was installed.

When boilers were tested in late July, they raised enough steam to blow the mill's whistle. Fred Haskins, construction superintendent, and R. F. Barker, operating superintendent, cut the first log on September 1 during a trial run on one side of the two band mill. By late September the mill was having daily "tune-up" runs; by late October it was running steadily. Only the small band head rig (a nine-foot band saw) was operational at the time as the large band head rig machinery was not completely installed.

There were plenty of logs to saw, as the logging crews had already landed 2 million board feet of logs in the log pond. Logging by rail started as soon as the Butte County Railroad reached company-owned timber for sale. Cedar, which had little demand as lumber, supplied the shingle mill.

After the lumber left the saws, it went via live rolls to the green chain for sorting according to grade. Underneath the green chain, on the east side of the mill, were ten 30-car tracks where the lumber was loaded directly on the railroad cars for shipment to Barber.

The two band mill had steam operated "shotgun" feeds to operate the carriage. The latter moved the log through the saw, with the "shotgun" whipping the carriage back to its starting point. The logs were held

Butte County Railroad No. 3 brings a three-car train into Stirling City just after completion of the railroad in April, 1904. Note the mainline still lacked ballast. (D. I. Corp. collection)

ber, with the initial logging off Log Spur No. 1 near the Doon Mill.

The other side of the mill was running by mid-October. At the same time, construction work was being rushed on a 216-foot extension on the back of the mill to provide space for the planers for finishing long timbers. It is interesting to note that the mill was designed to handle long timbers. It was capable of cutting timbers up to 60-feet long with the use of trailer carriage and timbers this long were cut and sized down on occasion during the mill's early years.

The planing mill was directly in back of the mill, the machinery being powered by the main line shaft from the Corliss engine. The sawmill was also equipped with lath and shingle mills. The lath mill was supplied with edgings to be run through it and shipped to Barber.

in position by sharply pointed spikes called "dogs." The "dogs" at this time were hand operated. The carriage setter rode the carriage strapped to his seat. It was not until years afterward that compressed air "dogs" were developed and it was no longer necessary for the setter to ride on the carriage.

The shotgun feed at Stirling City was fast: apparently somewhat faster than most of the shotgun feeds used in other mills. An excellent contemporary account of the "shotgun" has been related by early day lumberman Warren Ormsby:

> "Warren Ormsby had been working in a redwood mill on the coast in the summer of 1904 when he quit and went down to San Francisco. Looking, thereafter for a job, he went to a job office, Marion-Reedy, that specialized in construction placements. There

CENTRAL BUSINESS AREA OF STIRLING CITY

(Drawn from Map of Stirling City,
V. F. McCray, 1903)

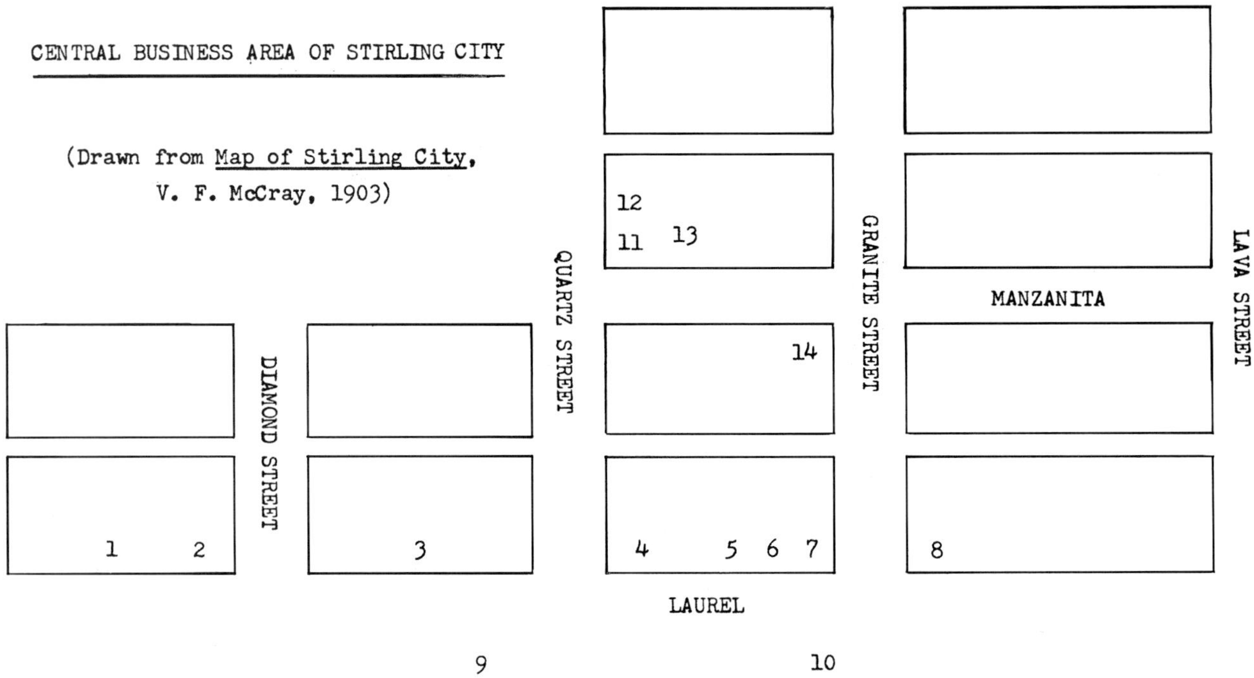

1. DANCE PAVILION
2. DANCE HALL*
3. RAYNOR HOTEL
4. STIRLING MERCANTILE CO.
5. NASH'S STORE AND U.S. POST OFFICE
6. POOL HALL
7. ROBE HOTEL
8. SPANGLER'S ROOMING HOUSE
9. STIRLING CITY BANK
10. NASH'S MOTION PICTURE HOUSE *
11. RED DEVIL SALOON
12. WHITE ANGEL
13. JAIL
14. JUSTICE OF THE PEACE

* Buildings in existence as of 1974. The Dance Hall was used as a rooming house after the fire in 1931. It is presently an antique store. The Motion Picture House has been used as the Community Hall since 1931.

K. Stephens, 1974

The Stirling City Bank was on Laurel Street near the mercantile company. (Mrs. Eda Seaman Scott collection)

were three jobs listed that he considered, one each at McCloud and Weed, pulling on the green chain and one at Stirling City, setting on the carriage. He paid Marion-Reedy $1.00 for the job and took the train to Chico.

"Taking the BCRR, he got into Stirling about noon but instead of presenting his job card to the boss, he just wondered over to the mill to look around. They were only running one side then and the steam 'shotgun' whipped the carriage back and forth so fast, with the sawyer turning the log while the carriage was in motion — flipping the dogs out as soon as the cut was finished, letting the log slide away from the blocks, using the nigger at the back-slack of the carriage — that it looked a lot more dangerous than the slower redwood mills where he had worked. The setter was strapped to the carriage and Ormsby heard he worked four hours on with one hour off for rest.

"Ormsby says he looked at it for a long time and said to himself, 'Nunh-Hunh!' He caught the lumber train back to Chico, got a job picking fruit and after making a road stake, ambled up to Portland, Oregon, to wait for the Lewis and Clark Exposition to open in 1905.

"Ormsby mentioned, too, that Diamond was paying by the day while it had been his experience in the redwoods to be paid by the month. This did not bother him nearly so much as that Damn Steam Shotgun. It was TOO FAST.[2]

The first construction at Barber was in February 1903, when work started on an inbound freight shed next to the Southern Pacific tracks. Almost at the same time another crew started building a four-stall roundhouse for the Butte County Railroad. This was Barber's first permanent structure.

Most of the building at Barber was delayed until the latter part of 1903, as Diamond intended to use its own lumber cut at the Stirling City mill for most of the construction. Other than the roundhouse, the only other permanent structure started was a large brick building to house the machine shop and foundry, which Diamond called the Engineering Department.

The company realized by October that the construction delays on the Butte County Railroad would prevent the railroad from being completed in time to transport the lumber for the buildings at Barber. The long-delayed construction work then started with the necessary lumber being purchased from the Sierra Lumber Company at Chico.

Crews were erecting numerous buildings at Barber by the spring of 1904. These structures included the powerhouse, dry lumber shed, dry kilns, storehouses, the sash door and box factory, general office building, and the Engineering Department.

The match factory was the last Barber facility built and also the last to begin production. There were several logical reasons for this. Diamond, over the

Top: The original building of the match factory shortly after completion in 1906.
Center: Delivery teams and yard personnel in front of the lumber shed at the first Chico retail yard.
Bottom: The main powerhouse at the Barber plant under construction in 1904. The 150-foot stack was an unsupported steel column with a platform at the top. (All D. I. Corp. collection)

Interior view of the Engineering Department bears out contemporary statements that it was a well-equipped machine shop. The machines were powered by a direct-line shaft running from the powerhouse. (Twenty Seven Years)

twenty year period prior to the development at Barber, had a small volume of match sales on the Pacific Coast – and thus no pressing reason to rush into production. A second and perhaps overriding reason was the conditions during the period for the manufacture of matches. The materials used, coupled with the intense heat of the Sacramento Valley during the summer months, made the manufacture of matches very unsafe. This was prior to the development of any type of air conditioning; until the Barber factory was air conditioned it was impossible to have year-around operation.

Plans for the match factory were approved in 1905 by Diamond's Board of Directors and construction started sometime during that year. The building was finished early in 1906 and the machinery installed a few months later. The first machine was in operation by March 29, but the factory did not start regular production until the cooler fall months.

The match factory's powerhouse was supplied with slabs and hogged fuel shipped to Barber over the Butte County Railroad from the sawmill. The fuel was also used to some extent in the main powerhouse at Barber, since the waste from the millwork department was not sufficient to supply the boilers.

Among the most interesting of Diamond's original facilities at Barber was the Engineering Department. The "department" was actually a large and well-equipped machine shop. It was a necessary investment for the Barber-Stirling City operations, since the shop required the capability to repair anything from a factory machine to a locomotive. This was because the nearest large commercial machine shops and foundries were in the San Francisco Bay area. This was more than 100 miles from Chico, and too far should any emergency arise. Then too, the company had experience with machine shops and foundries at other locations: one was the large Engineering Department at Diamond's Barberton (Ohio) plant. From their own experience, officials knew that the company could do its own work at a profit and even more important, at a minimum of delay. A well-equipped machine shop could also compete for outside business that could bring in profits.

James Chambers was master mechanic and supervised the Engineering Department: he had been hired by Mr. Clough from a position with the Tacoma & Eastern Railway. Most of the theoretical knowledge for the design of machinery was provided by Chambers or Fred W. Lane, who was Diamond's supervising western engineer.

Within a few years the Engineering Department built a large volume of business. Outside jobs included repair work on locomotives and railway equipment owned by regional lumber and mining companies, the manufacture and sale of a line of railway track tools, and dredger repair work. Jobs also included manufacturing, on a cash basis, anything that an inventor brought to them. Equipment designed by Lane and Chambers for Diamond's own operations was also selected for commercial marketing. This included the "California cut-off" saw, originally developed for the Barber box factory. Another was the "Barber" logging donkey, a double-drum logging donkey that was soon nicknamed the "Whistling Barber" from the sound of steam from its piston valves. Both the Barber donkey and the "Seattle" donkey were major improvements over the old vertical spool Dolbeer donkey. Barber donkeys were sold to a number of northern California lumber companies.

The Department had a few failures in machinery design: their biggest was the strange story of the Diamond Creeping-Grip tractor. This was a crawler tractor designed for agricultural use. It was one of the earliest applications of the crawler design, and may have been the first, predating the first tanks of World War I. Production started in 1913, with the Western Implement & Motor Company as sales agent.

The Creeping-Grip tractor was one of the largest tractors to be marketed for agriculture. The 10-ton behemoths had 72-horsepower engines, contrasting with 45 horsepower developed by the largest Holt tractor. They sold for $4,250 each. The only portion of the tractors not manufactured by the Engineering Department was the engine, which came from an outside supplier.

The initial production models, however, developed mechanical problems. The tubular radiator was unable to cool the engine, and core radiators were purchased outside. The most troublesome part of the tractors proved to be the continual breaking of the cast iron tread plates. Before these problems could be completely ironed out, the entire project abruptly floundered in the early months of 1914. Western Implement & Motor failed and Diamond secured a judgment against this firm. Two completed tractors were finally sold to pay the claim and the others were cut up for scrap. Both tractors sold went to farmers in the Chico area, and at least one of them worked in the rice fields for many years afterward.

Returning to the development of Barber, the company planned to develop a model town for employees around the plant facilities, with the community forming a suburb to the adjacent town of Chico. The plant name was extended to the proposed town. A reporter for the *San Francisco Chronicle* visited Barber during 1904 and wrote:

> The men who are handling the building of the town declare that they are following out the ideas of the model town of Pullman and profiting by the mistakes made there. A number of model homes costing from $800-$1,500 are already to be seen.[3]

In order to aid the development of the model town, some of Diamond's officials incorporated the Chico Investment Company on November 26, 1902. Fred M. Clough was the company's largest stockholder; all seven of the other stockholders were connected with Diamond. The company's purpose was to purchase acreage for subdivision into lots. The company made

One of the many locomotives owned by other companies that were repaired by the Engineering Department. Here former S. P. 4-4-0 No. 1486 sits next to the machine shop in 1907. (D. I. Corp. collection)

The "Whistling Barber" logging donkey manufactured by the Engineering Department. Nameplate shows this machine was manufactured in 1907. (D. I. Corp. collection)

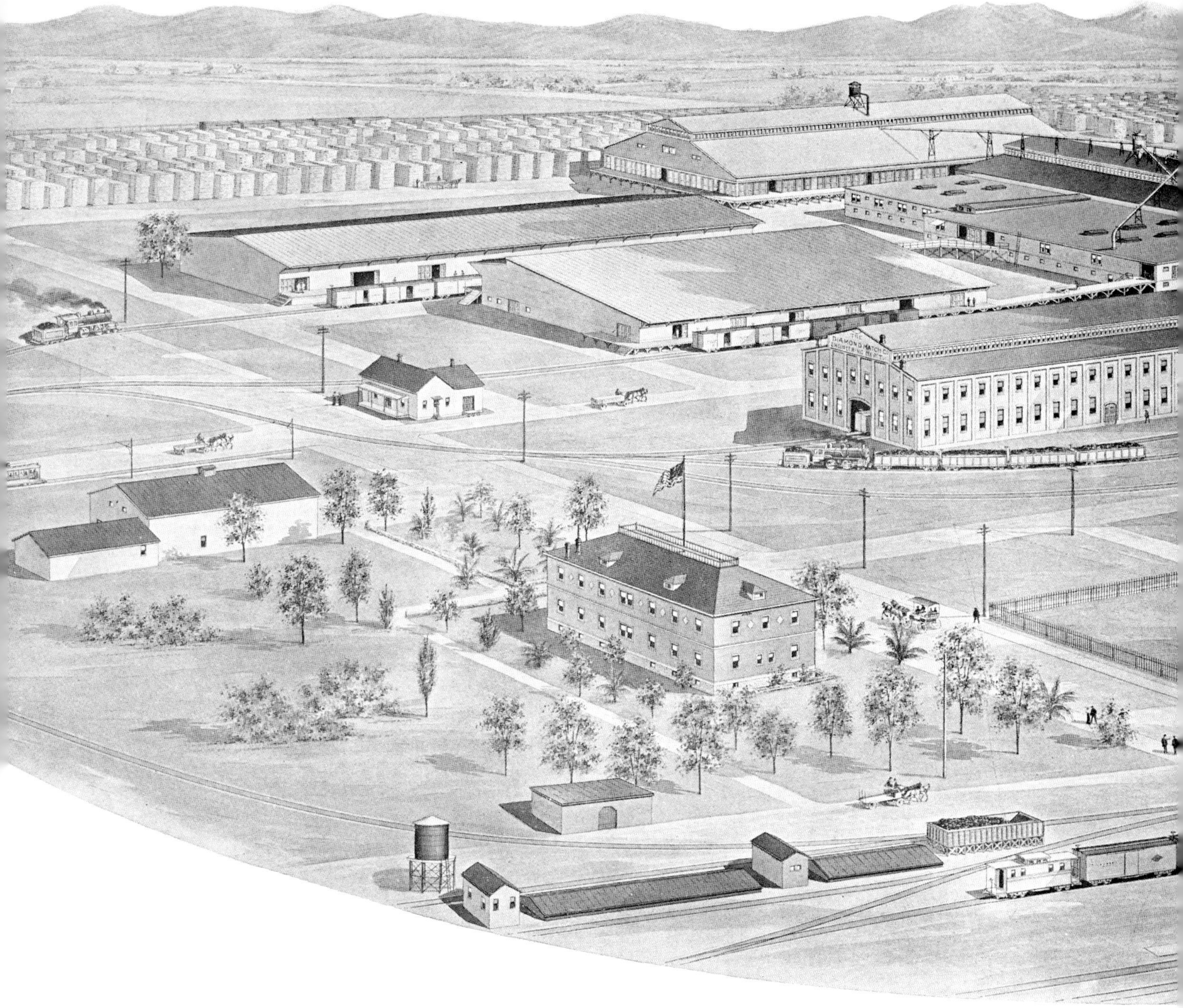

Panoramic view of the Barber plant after completion of the match factory in 1906. In front center was the Butte County Railroad's Barber depot. Behind it, on the right, was the match factory with a small warehouse behind it. The large brick building beyond was the Engineering Department machine shop and foundry. Behind the foundry was the main powerhouse. The building behind the machine shop with the sawdust pipes was the box factory; behind it was the sash and door factory. The two buildings to the left were warehouses. The two-story structure with the flag was the office. (Twenty Seven Years)

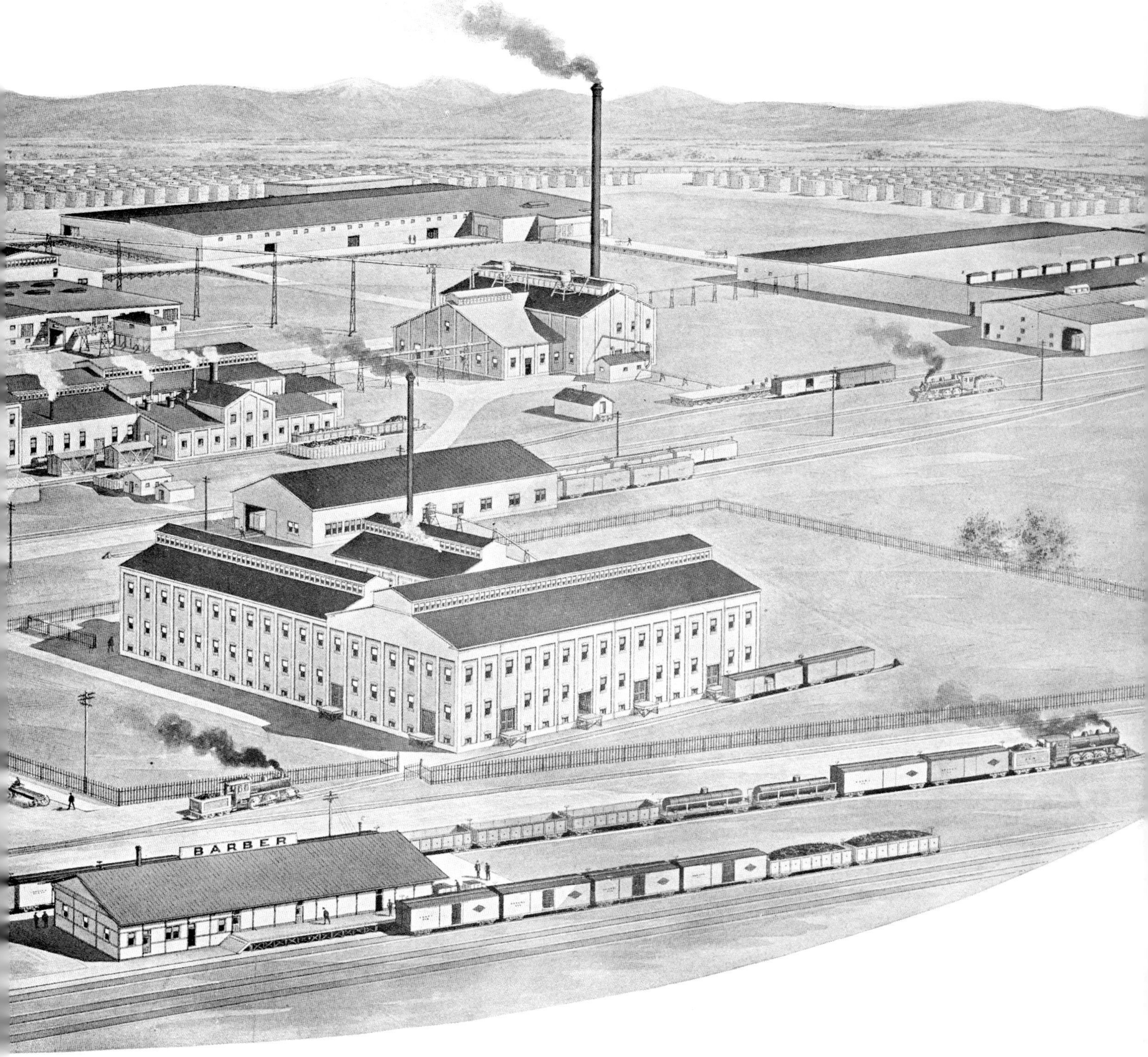
BARBER

The Creeping Grip tractor designed and manufactured by the Engineering Department was assembled in the machine shop (left). The original version with the unsuccessful tubular radiator is shown in the picture below left. (Both D. I. Corp. collection)

Apparently the same tractor is shown above after modification of its radiator. (D. I. Corp. collection)

The picture at left may be the other tractor, possibly modified for rice field service. (Mrs. Justa Farley collection)

Northern Electric train No. 6 pauses to pick up passengers near Fifth and Main in downtown Chico. (Kirby Harris collection)

three subdivisions totaling 449 lots between 1903 and 1905. This development, today part of Chico, extended south from Little Chico Creek to the present West 22nd Street, and west from Park Avenue to the east edge of the plant.

Many houses in the company's subdivision were built by the Chico Construction Company, a private venture of Fred M. Clough in partnership with W. A. Perley, Diamond's purchasing agent for the California operations. Perley later became Clough's son-in-law.

The largest and finest house in Barber undoubtedly was Mr. Clough's own home. At 1354 South Broadway, it was occupied by the Clough family until 1910.

The economic impact upon Chico with arrival of Diamond brought considerable growth. Before Diamond came, Chico was a small agricultural community whose largest industry was the Chico Division plant of Sierra Lumber Company. Diamond brought a much larger industrial payroll to Chico than Sierra's relatively small factory. The impact of Diamond's location is reflected by the records of the U. S. Bureau of the Census, which show Chico Township having a population of 4,739 in 1900. By the 1910 Census the population skyrocketed to 11,775 and Diamond was almost solely responsible for this boom.

Such a growing community could not continue spreading out in that bygone era of chiefly horse and buggy transportation without feeling the need for a streetcar line. Again, Mr. Clough and other company officials took the lead. Attorney F. C. Lusk, representing Clough, presented petitions for streetcar franchises to both the Butte County Board of Supervisors and the Board of Trustees of the City of Chico on December 7, 1903. The petitions outlined plans for a system serving all of Chico and extending into the suburbs, linking both to the Barber plant. The petitions resulted in the formation of the Chico Electric Railway Company, but Mr. Clough had sold his franchises prior to construction of the electric line. None of Diamond's officers were involved in ownership of the Chico Electric Railway Company after the franchises were sold, but there was some overlap in management. Diamond freely loaned management personnel and equipment during construction and Frank A. Clough, one of Fred Clough's sons, stayed on to be superintendent of the system. The company was sold to the Northern Electric Company in 1906, becoming part of its system. Service in Chico continued until World War II.

Northern Electric car No. 72, running westbound on the Fifth Street line, crosses the Southern Pacific line next to the S. P. depot. (James McK. Wright)

First Northern Electric depot at First and Main in 1910. Car No. 253, shown here, was built as a trailer, but rebuilt and motorized in 1910. (James McK. Wright)

Car No. 22 of the Chico Electric Railway Co., one of four cars owned by the streetcar line. (John Nopel collection)

CHAPTER FOUR

Expansion

DIAMOND'S first full season, that of 1905, found the mountain operations at Stirling City employing approximately 200 men in the mill, with an equal number in the woods. Another 100 busily extended the logging railroad. The 400 men involved in logging and sawmill production put in a hard summer's work, as existing records indicate the Stirling City mill ranked second in production among the mills in Northern California. The 78 million board feet of lumber produced by the McCloud plant of Scott & Van Arsdale put that company in first place.

Despite an average annual production of about 30 million board feet of lumber, Diamond made little profit on the sales due to a large plant investment. This was a subject of concern to Bryant & May, who proposed in 1905 to double production to more than 60 million board feet annually; this would then result in a profit to the partnership. Fortunately, Diamond never implemented such a policy over a period of years. If it had, Stirling City would have become an abandoned lumber town in the middle of cut-out timberlands.

Thus it was of major importance to the future of Diamond's operations when a tract of approximately 93,000 acres of timberlands was purchased. This tract comprised the entire remaining timberlands owned by the Sierra Lumber Company, and was acquired March 19, 1907 along with most of the latter firm's plant assets. Sierra's timberlands include most of what today is the largest portion of the Northern Tract of Diamond's 220,000-acre tree farm. The acquisition of Sierra was also important for other reasons. From an operational standpoint, it put Diamond for the first time into meter gauge railroad and flume operation. The seven retail yards Diamond acquired from Sierra were the foundation of the large chain of building material yards and stores that Diamond now operates in Northern and Central California.

The purchase price for Sierra broke down to $1,380,000 for timberlands and $1,720,000 for equipment and plant assets. The Colemans sold Sierra at the right time, since Diamond bought the company at the lumber market's peak. Lumber prices, climbing steadily from 1900 to 1905, jumped sharply in 1906 after the San Francisco earthquake and fire. In March 1907, a "silent panic" swept New York. The stock market, apparently on its way up after making a slow recovery, received a setback in August and then rallied. But the roof fell during October when the stock market collapsed, causing a nation-wide financial panic. The price of pine lumber fell virtually overnight to $7.00 M (per thousand feet) compared to its high during mid-1906 of $27.00 M. Despite this, Diamond's California profits were better for 1907 than for any previous year.

Diamond's acquisition of Sierra had little immediate effect on the latter firm's Red Bluff Division. Fire destroyed the Champion Mill on September 15, 1907, but Diamond rebuilt the former circular mill as a single band mill, and resumed production. Various improvements were made in 1908 to the Red Bluff plant and yard. Operations, however, continued virtually the same as under Sierra for the next four years. In 1911 the Red Bluff plant was abandoned and dismantled. The Champion Mill operated for another year and then it, too, was dismantled. A. E. Engebretsen, who had used the flume to ship lumber cut at his Judd Creek Mill, arranged with Diamond to continue fluming his cut to Red Bluff. This continued only a few months after he opened his mill in 1914. A long section of the flume collapsed west of Finley Lake. Engebretsen couldn't afford to rebuild it and Diamond

The best hotel in Stirling City was the Raynor Hotel. It boasted such early day conveniences as a barber shop, dining room, and hot running water. (Florence Styles collection)

Mount Saint Agnes Hospital was built on a hill at the south end of Laurel Street. It was operated by a Catholic Order, Sisters of Charity. (Florence Styles collection)

A peaceful winter scene looking south on Laurel. Nash's moving picture house is on the left, while Mount Saint Agnes Hospital can be seen on the hill in the distance. Below it is the roof of the depot. (Florence Styles collection)

wouldn't do the necessary repairs. He closed his mill permanently and this was the end of logging in the Northern Tract until 1952.

Sierra's Chico Division fared much differently under the new management. At the time Diamond acquired Sierra, the latter company was moving its West Branch Mill to a new location at Soda Springs. The work was only partially complete, forcing Diamond to make an immediate decision on the proposed new mill and the disposition of the temporary mill already at the site. An inspection by Mr. Clough resulted in the decision to operate the temporary mill for one season, as there was not sufficient timber in the Soda Springs area to justify operation of a new mill on the scale proposed by Sierra. (Local legend holds that the move was a bluff on the part of Sierra to force Diamond's interest in buying out the company. The mill proposed for Soda Springs would have had production almost as great as Diamond's Stirling City mill.) Rather than extending the flume to the temporary mill at Soda Springs, the company used teams to haul the lumber to the old mill site at West Branch for fluming to Chico. Sierra's meter gauge logging railroad at West Branch was abandoned and dismantled without being moved to Soda Springs.

At the end of the season the temporary mill was dismantled. Accessible portions of the flume were also dismantled and the lumber sold. Today the flume is only a memory to some of the old timers; remnants still cling to the canyon walls.

Sierra's Chico Division remanufacturing plant was a little more than a mile from Diamond's much larger and more modern plant at Barber. Diamond promptly closed Sierra's factory, dismantling the machinery and moving stacked lumber to Barber. The buildings, however, were retained for use as a veneer (plywood) plant.

On November 11, 1907, Diamond's Directors authorized Mr. Clough to purchase veneer machinery for the California operations. The veneer manufactured by the plant was to be used in the manufacture of doors, matchwood for Diamond's eastern plants, and box shook for the California citrus industry. The choice of a plant site undoubtedly was made because the former Sierra factory and planing mill provided existing buildings. Location of the veneer plant at Barber would have meant construction of a building. An added factor was Sierra's old 9th Street railroad spur providing a direct connection with the Southern Pacific. This permitted switching logs from Stirling City directly to the plant.

Installation of the veneer machinery started early in 1908 and the plant produced its first veneer on April

Stirling City from the south end of Laurel Street about 1906. The mercantile building dwarfs the other structures. (John Nopel collection)

A crew unloads logs at the veneer plant in 1908. The logs were shipped from Stirling City over the Butte County Railroad. (D. I. Corp. collection)

The saw filing room in the Stirling City sawmill about 1912. (Mrs. Florence Styles collection)

Panoramic publicity photo of the sawmill and lumber train in 1913. Note the Butte County Railroad box car behind No. 1. (D. I. Corp. collection)

4. This date places Diamond as the *first* manufacturer of plywood in California. Diamond distributed samples of the first day's production to business firms and Chico residents.

The veneer plant was only one phase of Diamond's program for almost complete utilization of every tree cut on its lands. In 1906 the company announced plans for building a large turpentine plant at Stirling City following visits by Fred Clough and Fred Haskins to similar facilities in the South. Plans called for initially building a pilot plant that would use slabs from the sawmill from which turpentine would be extracted. (Virtually all sawmills at the time discarded slabs as waste, or burned them for fuel.) The proposed extraction process would either be a charcoal carbonization or a steam "sweating" process.

Diamond's projected total utilization program also related to paper required for its match factories. Several paper-making experts reportedly arrived at Stirling City in 1907 to conduct tests in using wood fibers from sawmill slabs to manufacture paper. The experimental plant, apparently a ground-pulp mill, operated during the summer and was stated to be successful. Plans were then drawn for a $60,000 pulp and paper mill to be erected at Stirling City. The capacity initially was sixty tons daily. A hydro-electric plant, operating on Diamond's own water rights, would provide the power.

Diamond's early plans were far in advance of most companies of the same era. The proposed development of hydro-electric power, along with virtually complete utilization of every tree cut, showed a conception of a plan for the integration of all natural resources into its operations. Coupled to this were sound forestry practices. Diamond pioneered in forestry among lumber companies in California and very possibly on the entire West Coast. The firm had the first fulltime fire warden in California in 1903; the first commercial forester in the state, also in 1903, and, in approximately 1906, the first commercially-owned fire-lookout tower. Diamond's crews also included men trained to fight fires, and the company's early systematic care in logging practices were far better than any other lumber firm at the time. Much of this was due to the influence of Fred M. Clough.

Sierra Lumber Company's planing mill and box factory in Chico about 1900. Water tank on tower was for fire protection. (D. I. Corp. collection)

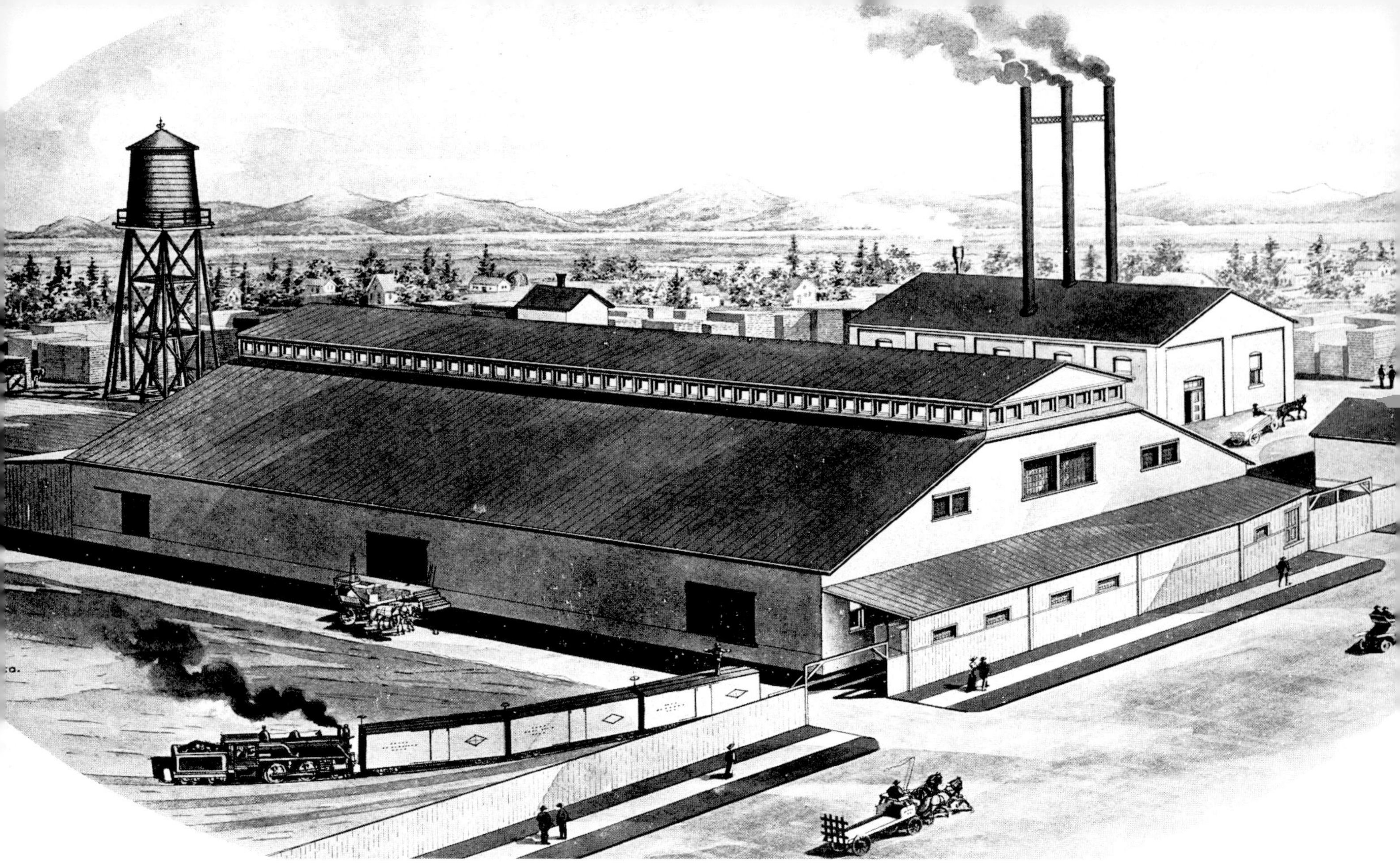

Artist's drawing of Sierra's Red Bluff plant in 1907 when Diamond acquired it. (Twenty Seven Years)

The veneer Plant in Chico just after it was placed in operation. This plant was the first plywood manufacturer in California. The artist apparently based his drawing of the steam engine on one of Butte County Railroad's 4-4-0's. (Twenty Seven Years)

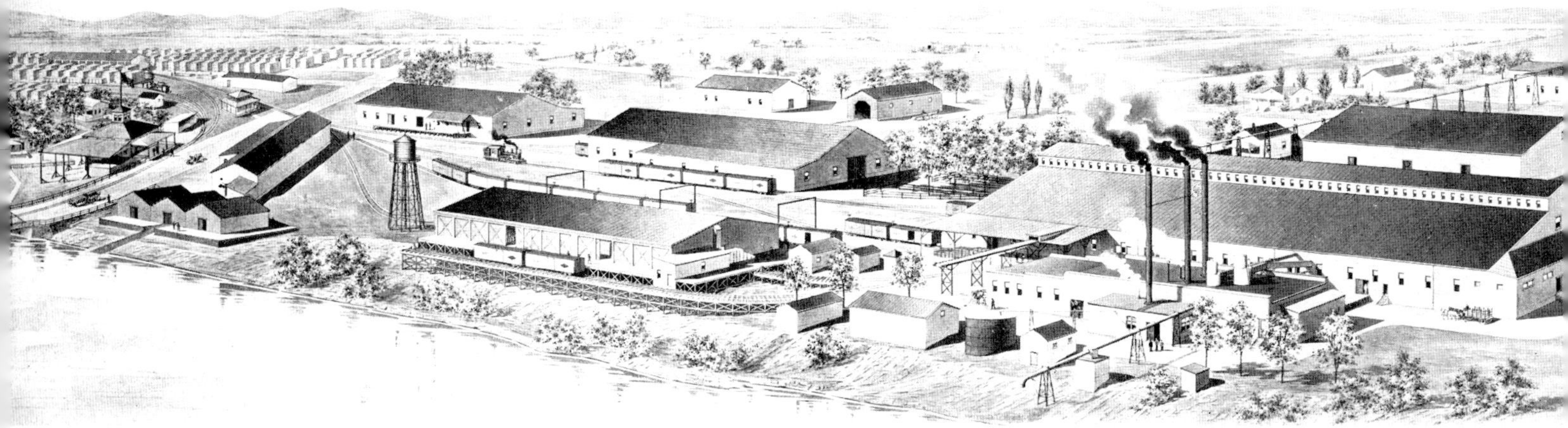

Former Sierra No. 3, the "Uncle Sam," was caught in the fire that destroyed the Champion Mill on Sept. 15, 1907. The engine was rebuilt and returned to service. (D. I. Corp collection)

A raging fire destroyed the Chico plant on Nov. 13, 1903. It was quickly rebuilt. (D. I. Corp. collection)

Sierra's West Branch mill at the head of the flume in Big Chico Canyon. (D.I. Corp. collection)

The sash and door factory at Barber about 1908. The factory was about three years old at the time, having manufactured the first door on Feb. 1, 1906. (Twenty Seven Years)

Artist's conception of the proposed $60,000 paper and box board mill at Stirling City. This was part of Diamond's planned near-total utilization of its forest resources without waste, a concept far advanced for its time. (Twenty Seven Years)

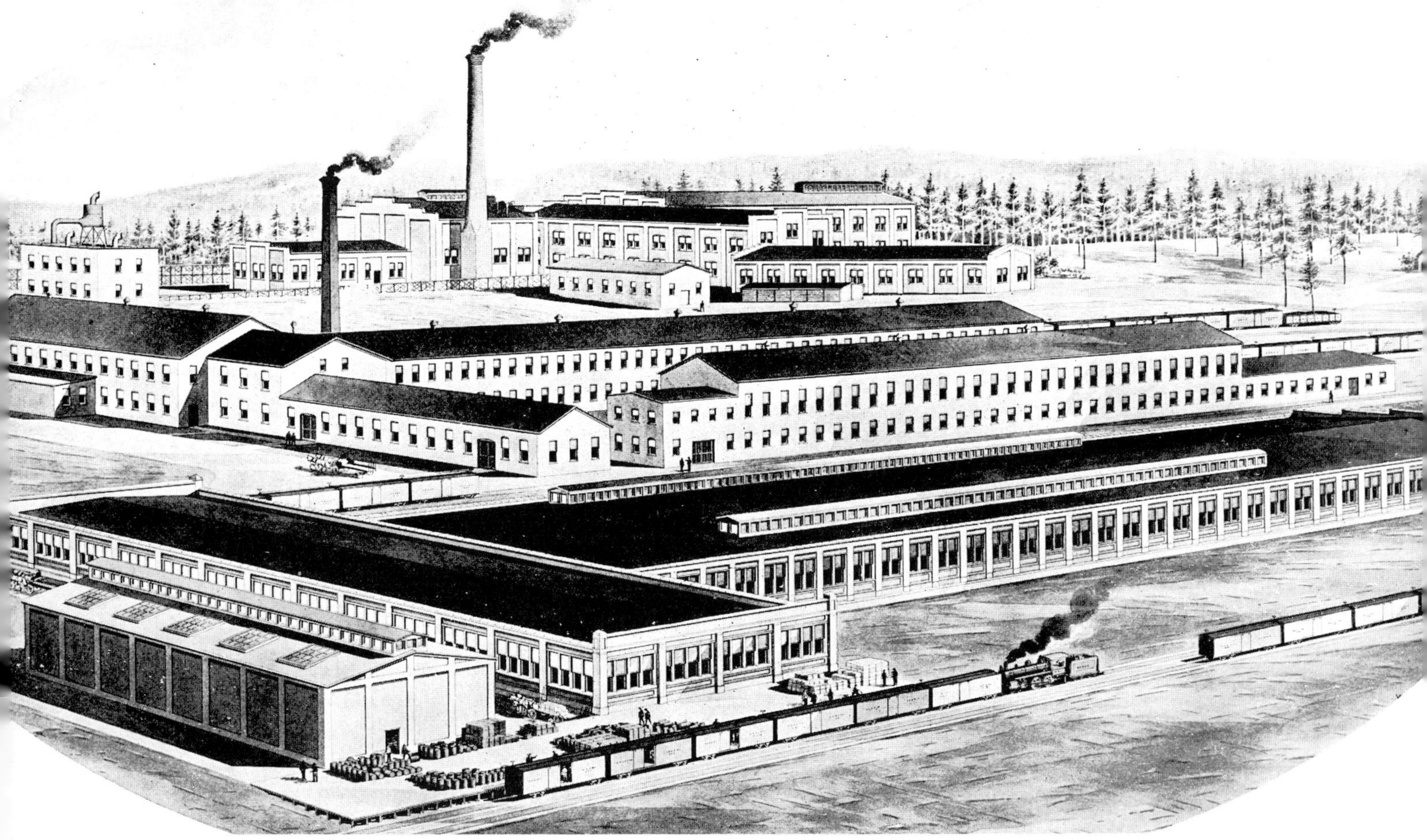

CHAPTER FIVE

In the Red

BRYANT & MAY had insisted for some time on a separate English company to handle the California operations for both firms. O. C. Barber, Diamond's president, opposed the idea. After a meeting of directors in February 1908, Diamond's treasurer, Edward R. Stettinius, wrote to Gilbert Bartholomew of Bryant & May with an offer that would solve the English company's objections. Diamond would either sell its half interest in the California operations, or buy Bryant & May's interest for the same terms: one-half of the total investment in California plus five per cent interest payable in twenty annual installments. Bryant & May didn't take long to decide to sell its interest, which Diamond's directors approved purchasing at their April meeting. The price was $2,250,000 – half of the total investment exclusive of the match factory. Despite their joint participation in the first five years, Bryant & May never used any California lumber or match plank.

From 1909 through 1915, most of Diamond's California logging production came from the Kimshew area east of Stirling City. Due to the exceedingly high cost of attempting to build a standard gauge logging railroad into the rugged area, the company decided to utilize the meter gauge equipment from Sierra and change gauges nine miles out of Stirling City. Headquarters for the Kimshew Show was established at Ramsey Bar, where the Stirling Mercantile Company operated a store and boarding houses. Kimshaw was the high point for employment for Diamond, with as many as 1,500 men working at Stirling City and in the camps. This high ball logging established a record in production during the 1911 season that wasn't exceeded until 1952, but the high expense of logging Kimshew's relatively poor timber came at a bad time for Diamond.

In 1909, the last of the top officials involved with starting the California operations left the company. O. C. Barber retired as president but retained a seat on the board of directors. His successor was Edward R. Stettinius, treasurer of the company. (This was Edward R. Stettinius, Sr., who subsequently served as Assistant Secretary of War in 1918-19. His son, Edward R. Stettinius, Jr., held various government positions beginning in 1941, culminating in the post of Secretary of State in 1944-45 and the first U. S. delegate to the United Nations during 1945-46.)

Shortly after becoming Diamond's president, Mr. Stettinius persuaded a brilliant 33-year-old engineer and naval architect, William Armstrong Fairburn, to join the company in the position of general superintendent as "the man to overhaul the organization." It was a timely choice in a critical period for the American match industry and for Diamond in particular. The manufacture of matches was extremely hazardous and dependent upon ideal weather conditions, subjecting employees to long layoffs while the factories closed during the hot summer months. The most critical problem was the occupational hazard of phosphorous necrosis, affecting employees of all match factories. The disease was caused by the fumes of the white phosphorus used in the manufacture of matches by all American companies.

Attempts to eliminate this serious hazard spanned many years. In Paris before the turn of the century, two chemists (H. Sevene and E. D. Cahen) developed a formula that substituted non-poisonous sesquisulphide of phosphorus for the deadly white phosphorus. The formula was adopted by the French government and in 1898 Sevene and Cahen received a United

Edward R. Stettinius succeeded O. C. Barber as president in 1909. (Twenty-Seven Years)

States patent for it. In 1900 the patent was sold to Diamond, which started developing the formula to produce matches the following year. Despite research on the formula by some of the best chemists in the following years, the company failed to produce a satisfactory "Strike-Anywhere" wooden match with the formula. The discredited formula was finally filed in the company archives.

By the time Mr. Fairburn joined the company, the problem of the toxic white phosphorus had become critical. Publicity over the hazards had finally resulted in public pressure upon Congress to ban the poisonous matches in favor of using the foreign-manufactured safety matches. Since the principle type of match manufactured by the American companies was the "Strike-Anywhere" wooden match, the American industry faced ruin.

During 1910 Fairburn and Henry Bentz designed, built, and placed in operation the first successful factory air conditioning system. Their system used cool water to lower the temperature and at the same time clean the air flow.

Once the ventilation-humidity control system was operational at the company's factories, Mr. Fairburn turned his attention to the match formulas. Early in his research, he discovered the discredited sesquisulphide of phosphorus formula in Diamond's archives. He started experimenting and late in 1910 discovered that the formula would work with chlorate of potash.

Representative John J. Esch in 1911 introduced a bill in the U. S. House of Representatives that placed a tax of two cents on white phosphorus matches, which in effect would tax them out of existence. On December 16, 1911, during a hearing on the Esch bill by the Ways and Means Committee, Mr. Fairburn announced that he had developed a method for producing a non-poisonous match that would be both safe and would satisfy the consumer. At the same time Mr. George Gordon Battle, general consul for The Diamond Match Company, stated to the committee that Diamond would grant all companies freedom from royalty payments and would issue licenses to any competitor for the use of the formula.

All American match companies, except for one, expressed appreciation to Diamond and indicated they were willing to change to the new formula. But one small company objected to the non-poisonous match and to using another company's patent. These objections upset the plans and resulted in an impasse over the enactment of the Esch bill. At this point Diamond was in a very advantageous position. If the bill failed, Diamond would be the only firm possessing a formula for a non-poisonous match. If the bill became law, the matches manufactured by its competitors would be taxed out of existence, giving Diamond an absolute monopoly in the manufacture of American matches.

Mr. Fairburn, however, believed that the humanitarian interests of health and safety within the industry and to the consumer were overriding factors, rather than the economic benefits the patent would bring to Diamond. He believed that it was the company's duty to make the discovery available to its rivals and persuaded both Mr. Barber and Mr. Stettinius to this view. After correspondence with President Taft, it was decided to remove all barriers to the use of the patent. On January 28, 1911, Diamond relinquished all rights to the patent and presented it gratis to the public.

The American Museum of Safety awarded its Louis Livingston Seaman Gold Medal for 1916 to Mr. Fairburn and the company for the elimination of an occupational disease.

William Armstrong Fairburn joined Diamond as general superintendent in 1909. In 1915 he succeeded Stettinius as company president. (Diamond Good Fellowship Bulletin)

Diamond's new management in 1909, however, had a problem more critical than phosphorus necrosis. This was the company's precarious financial condition. This was due partially to expenditures for the California development. Although far from the sole cause, the California operations were singled out by Fairburn and Stettinius for what, in retrospect, seemed an undue amount of blame. Neither had been involved in the planning and development of the California operations, and evidence indicates they had little patience for it, preferring the company's match-manufacturing area of operations.

The advanced utilization plans for Stirling City were a casualty of Diamond's financial problems. The proposed integrated turpentine plant, wood pulp mill, and paper factory was never built. Financial troubles first delayed and then caused complete abandonment of the plans. The veneer plant was also a casualty. It operated during 1908 and 1909 while having constant trouble with the glueline. This was a common problem in all veneer plants during this early period of the plywood industry. The veneer plant was idle in 1910, although sales were still being made of its already processed stock. The plant was dismantled in 1912. In terms of a profitable operation, the veneer plant can be called a failure, but it was significant in Diamond's early attempts at total utilization of its forest resources.

Another probable casualty of the financial problems was Fred. M. Clough, who was apparently pressured into resigning during 1910. He submitted the resignation to take effect on January 1, 1911. However, it was changed to take effect on September 1, 1910 — the day it was announced. The change reportedly was at Mr. Clough's request to enable him to "take a much needed rest." Undoubtedly this was not the real reason, despite the fact that it had been known in Chico that he had been considering retirement for the previous three years. He may have been anxious to go on to another project, since he had spent much of his life as an "organizer" who moved to another endeavor once he had "gotten things started." Balanced against this, however, must be the company's profit and loss situation. There is indication that the Stettinius-Fairburn management was not happy with Mr. Clough's grandiose manner of operations. From their viewpoint, it was not the time to be talking about a $60,000 pulp mill in California when the company was losing money.

Mr. Clough's resignation was received with regret both within the California operations of the company

Florance Armbruster, assistant treasurer, was placed in joint control of the California Division with Vernor Woolley in 1910. (Twenty Seven Years)

Diamond's second retail lumber yard at Chico, which replaced the earlier yard at the Barber plant. Before 1914, most of the yards were under the community rather than the corporate name. (Will Dean photo, Ira Brink collection)

and by many of the people in Chico. During his eight years in Chico, his community contacts brought him great esteem and respect. Both Mr. Clough and Diamond had very favorable press coverage on the coast. The community and his employees uniformly liked and respected him. A testimonial dinner, hosted by President Stettinius, was held for Mr. Clough on the night of September 1.*

Following Mr. Clough's resignation, Vernor S. Woolley (Assistant Manager) was appointed Acting Manager of manufacturing and outside operations, and Assistant Treasurer Florance Armbruster was placed in charge of inside operations. Apparently there was a short power struggle between the two men and on December 8 it was announced that Woolley had been named the new general manager.

Woolley, approximately 47 when he became general manager, had been with Diamond since 1906. He became the general superintendent at Chico in 1907, with jurisdiction over lumbering at Red Bluff, Lyonsville, Chico, and Stirling City. He was a colorful individual and was described by W. H. Hutchinson in 1957 as:

> ". . . a two-fisted drinker and a two-fisted fighter who would jump into the middle of a loggers brawl in a saloon and make the best of it. He knew how to handle men and his men loved him; even today they speak of him with great warmth. He was a square shooter, a rather flamboyant personality, and his social activities made news . . ."[1]

During his term as general manager, Woolley's residence was the former home of Mr. Clough, which the latter sold to Diamond upon his retirement. Woolley was troubled during the period he was general manager by the unsettled lumber market, the high cost of meter gauge logging, continued financial problems of the California operations – and finally the shock produced by the Jay Price timber cruise.

The top management of Diamond in 1912 began to realize that the stumpage estimates for the Kimshew area were in excess of the profitable timber that could be logged. Vernor Woolley and his subordinates reportedly knew at least a year earlier, but the New York office wouldn't accept their reports.

The first accurate "cruise" (or survey) of Diamond timberlands started in 1913. Jay Price, a recent civil engineering graduate of the University of California, headed the party. The cruise took two years, and on May 12, 1915, Mr. Fairburn learned the bad news in the report while on an inspection trip at Chico. The Price cruise showed a total of 920 million feet of standing timber on Diamond's lands. This figure had a shocking impact on Diamond officials because it was substantially under earlier reports. The shock can be realized when the Price cruise figure is compared with the statement in *Twenty-Seven Years,* a booklet issued

The Yuba City yard was one of the larger Diamond yards and used the corporate name. (Will Dean photo, Ira Brink collection)

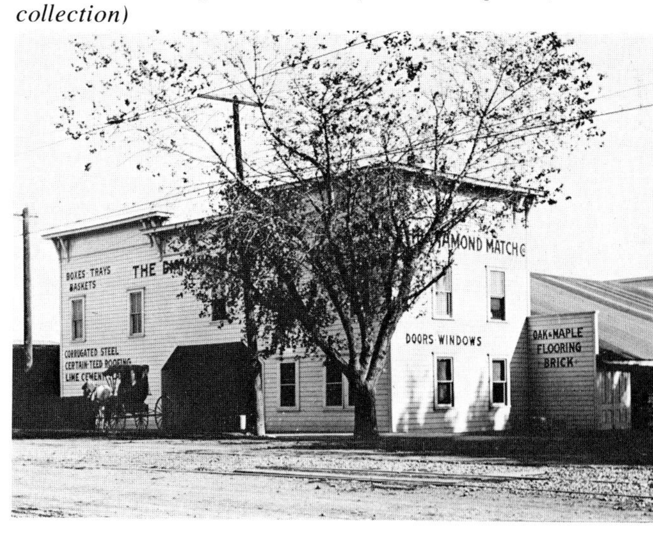

*At the time Mr. Clough resigned, his title was Vice President and Pacific Coast Manager.

After the Cloughs left Chico they resided in Oakland until the 1920's, where Mr. Clough was in part responsible for founding the Pacific Coast Shredded Wheat Company. It was this company that introduced shredded wheat to the West Coast. Mr. Clough retired from active business life in 1922 and the family moved to Los Angeles, where he passed away during the early 1930's.

Above: The Gridley retail yard was typical of the smaller yards in the chain. (Will Dean photo, Ira Brink collection)
Left: H. H. Gallagher, Willows retail yard manager, posed with the new Sterling Truck assigned to his yard in 1919. This truck and three identical ones were Diamond's first trucks for retail yards. (D. I. Corp. collection)

The retail yards marketed wooden silos to farms throughout the Sacramento Valley. This one was built at Lincoln in 1914. (D. I. Corp. collection)

The Woodland retail yard office in 1914. The man on right is believed to be the late Ira E. Brink, then manager of the Woodland yard. In 1923 he became manager of the retail yard system. (Will Dean photo, Ira Brink collection)

in 1908 by Diamond to its stockholders. According to the booklet, Diamond's California timberlands *by conservative estimate* contained approximately 3.5 *billion* feet of standing timber.

A total of 800 million feet of this figure supposedly was in the Kimshew area alone. The start of logging brought the discovery that this virgin stand contained a high percentage of over-ripe timber. This meant low grade lumber, some of which couldn't even be sold.

Price's greatly reduced figure does not mean earlier cruises were "padded." The Price cruise, on orders of Vernor Woolley, included only "operable" trees tallied one hundred per cent. Snags were omitted and the cedar almost ignored, while the tally on the burned trees and the fir was scaled down. In this way the cruise was a more accurate reflection of the value of the timber after it had been sawed into lumber.

The decision for a general cutback in the California operations had already been made by the time the results of the Price timber cruise were known. The basis for this decision was the theory that the losses could be reduced by a drastic curtailment. These losses, plus the shock of the Price cruise, were undoubtedly the reason for Diamond staying with a relatively low production rate throughout the rising lumber market of World War I. Diamond was learning a very bitter lesson in lumber industry economics: estimated stumpage on the timberlands did not equal logs at the sawmills or lumber that was marketable. Diamond, however, learned the basic lessons of lumber industry economics and remained in business in California.

Several reasons for their success can be identified. One very basic factor was the acquisition of Sierra's seven retail lumber yards when they purchased this company in 1907. Diamond wasn't interested initially in operating the seven yards, despite the fact they already had a retail yard in Chico and they tried to sell the Sierra yards. Before they managed to sell the yards, they learned a lumber industry lesson apparent years earlier to Sierra. This was the advantage of vertical integration in owning retail lumber yards to serve as outlets at retail prices for the sawmill production. The yards were retained and in the 1920's a marked expansion started. Gradually Diamond's operations took on the same appearance as Sierra's in its later years. Additional years of losses taught Diamond to not cut much more than what could be retailed through the yards.

CHAPTER SIX

Under the Slow Bell

PREPARATIONS were well under way at Barber and Stirling City by mid-1915 for the general cutback. Fairburn and the directors accepted the offer of Frank Thatcher, logging superintendent, to operate on contract all of the company's Stirling City operations. The millwork and engineering departments at Barber were closed. By August there were only twelve men on the entire Barber payroll, exclusive of the match factory, and most of these employees were either watchmen or farmers. What little millwork activity that remained was moved to a small building at the corner of West First and Orange streets in Chico, one block from the Chico retail yard. The lumber department's office at Barber was closed and the few remaining employees moved to a new office on the second floor of the building housing the Chico retail yard office.

The directors had also decided not to renew the lease on the Butte County Railroad, turning it over to the Southern Pacific for operation effective December 1, 1915.

Construction work started at Stirling City on a lumber piling yard adjacent to the mill. Lumber was to be first piled and dried to avoid shipping it green over the Southern Pacific, as had been the practice during operation of the Butte County Railroad. The reason for the change was that a higher rate was charged for the green lumber.

Contrasting the curtailment of the other operations, both the match factory and the retail yard operations were greatly expanded. The match factory was virtually doubled in size and its production went up accordingly. At the request of the match department employees, the empty lumber department office building was "given" to their unit. They renovated it into one of the finest employee clubhouses in the nation. The facility was dedicated October 18, 1916, as "Fairburn Hall," honoring William Fairburn. He had approved use of the building as a social hall and drew the plans for the renovation.

Vernor S. Woolley was named manager of lands at the end of the 1915 season, but he soon resigned. When he left, the California operations were divided into two segments: the Stirling City logging and sawmill activities were under contract to Frank M. Thatcher, while the valley operations at Chico and the retail yards were headed by William B. "Will" Dean. In theory it was two segments, but in practice Mr. Dean ran the whole show as Thatcher was required by his contract to cut according to what could be sold through the retail yards under Dean. What little wasn't sold by the yards went to the California Sugar and White Pine Agency for marketing.

Diamond's top management in New York also changed in 1915. Mr. Fairburn became acting president on January 28, when Mr. Stettinius took a leave of absence to join J. P. Morgan & Company as advisor on the purchase of war supplies for Great Britain. Mr. Stettinius resigned from Diamond in September, realizing then that his work would result in a long absence from the company. Mr. Fairburn was elected President of the Diamond Match Company on September 24 and filled the position with distinction until his death in 1947.

Diamond's lumber production in California remained "under the slow bell" all during the World War I period, but the sawmill cut was a little higher each year to satisfy the growing demands of the retail yard chain. This growing demand of the yard sales within three years resulted in the newly built millwork plant in Chico becoming inadequate for production needs. The

*After the cutback of 1915, Diamond's millwork plant was in this small building at the corner of First and Orange Streets in Chico. The building is still standing today, although Diamond vacated it in 1920. (*Diamond Good Fellowship Bulletin*)*

virtual abandonment of the Barber facilities ended in November 1919, when the company started rebuilding the millwork factory there. Production started early in 1920 and the small Chico millwork facility was closed. The millwork plant at Barber was the only portion of the plant that was reactivated. Neither the engineering department or the sash and door factory was ever reopened.

The cost of production at Stirling City went up steadily during Thatcher's contract. This didn't affect him, as he was operating on a cost-plus contract of 15 cents per thousand profit, negotiated in 1918. The change came because he lost money during the 1916 season and made very little profit in 1917. Part of the rising cost after 1920 is explained by logging railroad extensions into new logging areas near Inskip, where construction costs ran as high as $25,000 per mile.

The continual rise in the cost of the contract logging, however, resulted in the termination of the agreement. Mr. Fairburn visited Stirling City in 1922, noting many inefficiencies. The president decided it was time for a change: the contract arrangement ended at the conclusion of the 1922 season.

William Dean was appointed general manager of the California Division at the close of the season, getting the formal title for duties he was performing. Logging Engineer Frank A. "Comp" Compton became superintendent of logging and sawmill operations at Stirling City. "Comp" remained in charge for five years, being succeeded in 1928 by Richard A "Dick" Colgan, company forester. Again, it was a story of rising production costs on a falling lumber market.

Diamond's fire losses in its timberlands, heaviest following World War I, resulted in an average annual

Fairburn Hall in June, 1916. Formerly the office building at Barber, it was given to the employees as a social hall after it was vacated during the 1915 cutback. (H. C. Tibbitts photo, D. I. Corp. collection)

Superintendent Richard A. "Dick" Colgan at Stirling city in 1929. (Isabel Millington)

Will B. Dean was appointed general manager of the California Lumber Division in 1922. (Leta Robbie collection)

The match factory as seen from the Southern Pacific tracks in 1916. The building is little changed today. (H. C. Tibbitts photo, D. I. Corp. collection)

loss of 5,677 acres. But this was changed by an event in 1929. At a meeting of the State Board of Forestry in March, Director Fred Stevenot proposed development of a statewide forest fire protection plan which would include equipment and men. He further suggested that the board select a demonstration area to prove the merits of California having a formal forest fire protection plan. Deputy State Forester William B. Rider proposed that Butte County should be the demonstration area on the basis that "the Diamond Match Company could be depended upon to cooperate."[1] The Board passed a resolution for a joint project of fire protection in a concentrated area between a national forest and the adjacent county having state fire protection. This proposal became reality with formation of the North Butte Fire Protection District on December 31, 1929. The four parties to the agreement – Diamond Match Company, the State Division of Forestry, the U. S. Forest Service, and Butte County – agreed to furnish protection equal to three cents per acre of its lands. This amount, more than the actual cost of fighting fires, made the North Butte Fire Protection District effective; in 1930 the jurisdiction hired the *first* full-time forest fire fighting crew in California history. Diamond's fire losses dropped sharply with formation of the district, and by mid-1931 Dick Colgan reported to the State Board of Forestry that the district's efforts were successful. The district served as a pilot program for a fire protection plan for California, and remained in operation until 1947 when it was liquidated, with its duties assumed by the California Division of Forestry.

Half of a six-section panorama of the Barber plant in June, 1916. By then it was being called the Chico plant. From left: warehouse, box factory (behind), powerhouse, engineering

The Apiary Department occupied the old main powerhouse at the Chico plant after 1920. (D. I. Corp. collection)

The district's formation did not prevent Diamond from suffering, two years after the agreement, one of its worst blows from fire. It was on April 20, 1931, during an unusual spring drought that the real disaster occurred. About 11:20 a.m. a fire started under the stairway adjoining the Stirling Mercantile Company building. The fire, apparently ignited by a discarded cigarette, quickly flared up and spread so fast that tenants of various business buildings left their possessions and fled the burning buildings. Dick Colgan was

department, match factory warehouse, and the match factory. (H. C. Tibbitts photo, D. I. Corp. collection)

working at his desk in the Diamond Match Company office when the alarm sounded. Rushing next door to the Stirling Mercantile Company, he found the store's rear portion in flames. He returned to his office, grabbed an armload of records, carried them across the street, and laid them down. Ward Bennett, Diamond's bookkeeper, followed him out of the office carrying the payroll, which had just arrived that morning. Both men returned to the office to save more records but it was so hot they couldn't enter.

For the next three hours, a force of 250 men battled the raging fire. The strong northwest wind saved the mill and lumber piles, but flames virtually destroyed the historic Sierra Nevada lumber town. Many buildings erected when the town started were engulfed. Three square blocks, including ten business buildings and the homes of eighteen families, lay in ruins. The total loss hit $100,000, but the fire fighters work was effective as they saved property valued at more than $9 million. After the fire was contained, Diamond sent trucks to Chico to haul back supplies; a temporary store was opened in a vacant building. Rebuilding started immediately, but many of the original structures were never replaced, or were rebuilt in reduced size. The mercantile store was rebuilt on the same site, but a one-story building replaced the three-story landmark. Diamond never again operated the store itself and started leasing it to individuals. Although saved from total destruction, Stirling City was never the same after the fire.

Stirling City's population decline during the 1930's was more because of company operations than to the fire. Diamond, unlike many lumber companies, forced to close during the depression, continued lumber operations even though at a reduced level. It is to the credit of the company that every effort was made to keep long-time employees – especially those with families – working during the depression. Several factors made continued operation possible at Stirling City. Diamond's business was not limited to lumber nor were operations confined to Stirling City and Chico, in marked contrast to many lumber firms. Diamond's di-

A new office building was built for the Chico plant's resumption of operations in 1920. (Leta Robbie collection)

The Diamond Match Co. band was organized at the Chico plant in 1935 when it was discovered there were thirty-two musicians working at the plant. The band played at company picnics, ball games, and local concerts. (Diamond Good Fellowship Bulletin)

versification with other manufacturing facilities and products permitted it to withstand the lumber market's collapse. The company was in a strong financial position to weather the depression's effects and continued in business on a curtailed basis.

The National Recovery Administration code, effective in August 1933, allowed Diamond a quota of 30,684,000 board feet per year for the California Lumber Division. This was a reduction of approximately 5 million board feet under what the company had determined to be its minimum needs for the retail yard chain and other uses. On June 15, 1934, Diamond applied to the Lumber Code Authority for a sustained yield certificate. After investigating Diamond's forestry practices, the Western Pine Association, acting under provisions of the Lumber Code Authority, issued a certificate allowing an additional ten per cent over the lumber cut established by the earlier quota. The effective date for the request was October 1, 1934, making it the *first* sustained yield certificate granted by a Federal agency to a Pacific Coast lumber company.

The second West Coast firm to be certified by the Western Pine Association was Fruit Growers Supply Company at Susanville. Both the Michigan-California and Weyerhaeuser lumber companies apparently had applications pending when the N.R.A. was declared unconstitutional.

General Manager William B. Dean retired in May 1937, after completing fifty years in the lumber industry. His successor was John R. Gray, general manager of the Northwest Timber Division with headquarters at Spokane, Washington. Gray was no stranger to the California Division, having been the last superintendent of the Butte County Railroad. He continued in his position at Spokane, heading both divisions with the title of Pacific Coast Manager. Managing the California Division and reporting to Gray was a "troika" that assumed increased responsibilities following Dean's retirement. Dick Colgan became production manager, replacing Dean in authority over timberlands and production. The troika's other two members were Ira E. Brink, retail manager, and M. G. Atchison, auditor.

Stirling City Sawmill - 1936

Log wash on slip.

Log on carriage ready for band saw.

Rear of the mill, showing the edger in the center, trimmer on right.

Trimmer. Operator's platform in foreground.

Upper band saw wheel. (Five photos, Richard Colgan collection)

Laurel Street, Stirling City, in the late 1920's, looking south. Building at left was the theater. (Norman Hubbard)

Granite Street in the late 1920's, an assortment of houses, stores, and offices. (Norman Hubbard)

Laurel Street before the fire. The great size of the mercantile building is evident in this February, 1914, picture taken after a snowstorm. (Vera Hedge)

About the same view after the fire. Kilpatric and Sons store is now on the site of the old mercantile building. (Eastman Studio photo, Vera Hedge collection)

Stirling City in the late 1930's from the south end of Laurel Street. (Eastman Studio photo, Vera Hedge collection)

The same street photographed from the north end. (Eastman Studio photo, Vera Hedge collection)

Dry lumber yard south of the Stirling City sawmill in the mid-1920's.

Lumber was placed on two-wheeled carts. A bolster was then placed under the front of the load, and a horse hitched to the four-wheel rig for hauling. George Hubbard stands on this lumber cart, with the future Mrs. Norman Hubbard in the background.

Loading dry lumber on railroad cars on the "brawl" of the mill. (Three photos by Norman Hubbard)

Diamond started trucking lumber directly from Stirling City in 1932 using the fleet of trucks based at Chico. The most unusual aspect was the name – the VV Transportation System – from the "Vinegar Valley Railroad," a kind of toonerville trolley. This started as an office joke over a late delivery and the name stuck. The view above shows the truck fleet in 1935 lined up for an industrial parade. The picture below is of the drivers and twenty-seven trucks in 1940.

Part of the present truck fleet, which totals twenty rigs. Most are diesel powered. (Three photos, D. I. Corp. collection)

CHAPTER SEVEN

End of an Era

UNLIKE THE LOW production that characterized Diamond's operations during World War I, a heavy output came in World War II. This high production was particularly true at the Chico plant, where the original millwork department's production records were exceeded by a smaller plant. It supplied most of the millwork demands for the contractors that built more than 3,000 homes in the San Francisco Bay area cities of Richmond, Napa, and San Carlos. Doors alone were manufactured at the rate of between 2,000 and 3,000 per day.

The management in California changed at the end of the war. John R. Gray passed away in 1945 and was replaced as Pacific Coast manager by Joseph E. Duffey from the New York office. The same year Dick Colgan resigned as Diamond's production manager to become executive secretary of the National Lumber Manufacturers Association in Washington, D.C. M. G. Atchison served as acting manager of the California Lumber Division for nearly four years. Earl Bechard, who was from Diamond's Minnesota operations, became manager in 1949.

In the early 1950's the long dream of a mill in the Northern Tract of Diamond's Tree Farm became a reality. This was the first mill in the Northern Tract since the abandonment of the Champion Mill in 1912, and in the following forty years numerous proposals for such a facility were made without results. Diamond officials planned that the new mill would salvage overmature and beetle-attacked pine not accessible to the Stirling City mill. The site selected for the mill was Lyman Springs, 34 miles east of Red Bluff and only a few miles from the abandoned site of Lyonsville. Construction started on the mill and five miles of access road in October, 1952, and the first log was cut by the headrig in April, 1953.

The Lyman Springs mill, however, had a relatively short life. By the mid-1950's Diamond decided that neither the new Lyman Springs sawmill nor the aging Stirling City mill was adequate to meet the company's future needs. Very little logging had been carried on in the Northern Tract of the 230,000 acre tree farm since the Champion Mill had closed. Conversely, logging for the Stirling City mill had been carried on in the Southern Tract since 1904. (The timber of the Northern Tract was not accessible to the Stirling City mill without a long haul.) For this reason it was decided to locate the new forest products factory in an area where the timber of the Northern Tract would be accessible. Diamond announced in 1956 it was returning a major operation to Red Bluff, where it had closed the Sierra plant nearly a half century before.

The Del E. Webb Construction Company of Los Angeles and Phoenix and the Southwestern Construction Company of Los Angeles started work on the new plant in July 1956. Only thirteen months after the initial pouring of the foundations, the first carload of 300,000 molded pulp egg cartons left the plant on August 31, 1957, via the Southern Pacific Railroad for Diamond's General Package Division at Los Angeles. During the relatively short construction period the contractors completed a $14 million project on a 400-acre plant site. The Red Bluff Integrated Forest Products Center was formally dedicated on October 10, 1958, with state and local dignitaries present for the ceremonies. Capacity of the modern two-band sawmill is about 100 million board of feet of lumber annually, with a 30,000 ton annual molded pulp production.

The new forest products plant replaced both the

Aerial view of the sawmill complex during the 1940's. The outside braced building was the fuel bin. Next to it was the powerhouse with its five stacks. Behind the powerhouse and the adjacent sawdust burner was the machine shop, and behind it stood the enginehouse for the logging railroad. The small building at the lower right was the planing mill. (F. G. Lust photo, D. I. Corp. collection)

J. W. "Bill" Bloxham, stationary engineer, inside the Stirling City sawmill powerhouse about 1950. (Josephine Bloxham)

Lyman Springs and Stirling City sawmills. The end came for the Stirling City mill on January 31, 1958, when the last log went up the slip-way at 3:00 p.m. The mill closed shortly afterward. The *Chico Enterprise-Record* noted the closing of the Stirling City mill and called it the end of "one of the most historic and best known industrial establishments in the Northern California area,"[1] During its 54 years of operation, the sawmill produced 1,809,000,000 board feet of lumber. Most of this output was pine, which totaled 65 per cent of either ponderosa or white pine. The remaining 30 per cent was white or Douglas fir and the balance cedar and other woods.

There had been other closings of the mill. Each year since 1904 there had been an annual shutdown for winter. But this was the last gasp. As the mill went through its final paces that cold January day, most of the town's residents and a number of retired employees sadly watched the end of a part of what had been their daily life.

The Lyman Springs sawmill just after completion in 1953. This mill had the shortest operational span of any of Diamond's California sawmills. (D. I. Corp. collection)

Many employees at Stirling City were offered transfers to Red Bluff, and gradually a large portion of the town's population began to move. Officials of the Diamond Gardner Corporation* helped ease difficulties of the move. Other veteran employees, including Logging Superintendent Dana Baily and Sawmill Superintendent Edward Mason, retired. Although the mill was closed, silent, and slated to be dismantled, Stirling City did not become a ghost town. Diamond continued to contract some selective logging in the Southern Tract and the logging contractor made his headquarters in town. A number of the loggers lived in Stirling City to be close to their jobs.

The Stirling City sawmill was sold late in 1958 for dismantling to Pollak Steel Supply. The Chico-based company started scrapping the big mill the following spring, but by a strange turn of events they didn't get much salvage. At 2:00 a.m. on April 10, the town fire siren started blowing and the townspeople awoke to find the mill on fire. Within minutes the entire mill area was engulfed in a raging inferno and the roof caved in. A northward wind, similar to the conditions that saved the mill 28 years earlier when most of the town burned, kept flying embers and boards out of the town. The intense heat, however, started a number of yard and roof fires. Fire units of the California Division of Forestry fought the fire, which could be clearly seen 12 miles away in Paradise. All through the height of the fire, the log slip-way stood silhouetted against the raging flames, stubbornly refusing to be destroyed. The fire was finally controlled after covering ten acres, burning the entire mill area and spreading into the surrounding timberland. The cause of the fire was not definitely determined. Mr. Pollak, who lost some of his machinery in the fire, stated the entire mill area had been carefully checked and all hazardous areas watered down with no sign of fire at 4:30 p.m. Officials of Diamond Gardner believed that a cigarette or spark from scrapping operations lay in the sawdust somewhere in the vast mill area and was fanned by the wind that night. Despite the official verdict, there was another theory of suicide by fire. As the *Chico Enterprise-Record* put it, ". . . Many veteran woodsmen and millhands felt the old queen of the mill town preferred this spectacular end."[2]

For the next few years Stirling City was quiet without the customary sounds of a sawmill. Then in August 1963, Diamond announced it was returning a sawmill operation to Stirling City. The new mill, placed in operation during March 1964, was a small electrically-powered sawmill designed for cutting immature timber. The immature timber, which needed to be thinned out before it died, was too small for regular lumber production. The main product of the mill was

*Diamond Match Company merged with Gardner Board and Carton Company on October 31, 1957. The latter firm's business was organized as the Gardner Division. At the same time, the corporate name was changed to Diamond Gardner Corporation, as the old name no longer reflected the diversified business of the corporation and emphasized only one part of the product line.

The name Diamond Gardner lasted only two years until another merger. On September 28, 1959, Diamond Gardner Corporation and United States Printing and Lithograph Company merged to form the Diamond National Corporation. In 1964 Diamond was organized into a holding company and named Diamond International Corporation.

Ruins of the Stirling City sawmill after the fire. Only the machine shop was left standing. (D. I. Corp. collection)

eight foot two-by-four studs for use in housing construction plus a by-product of wood chips. Production of the compact mill was 40,000 board feet per eight hour shift and it employed approximately 24 men.

It was expected that the sawmill would remain in operation for a decade until all immature timber was thinned. That prediction proved to be accurate: the sawmill was closed in May 1974, and was torn down shortly afterward. At the same time the company announced that all logging would cease in the Southern Tract for some years in the future.

Stirling City will undoubtedly survive as a town. Its residents, however, will have a variety of occupations and it could gain some attraction as a retirement community — which has caused nearby Paradise to grow — as well as a location for summer homes. At some future time it may again have another large sawmill. When the timber in the Southern Tract matures, transportation costs could be saved by locating the sawmill at Stirling City.

A year after the closing of the stud mill a portion of the Chico plant suffered the same fate. Citing changing economic conditions, Diamond announced on October 9, 1975, the "immediate" closure of the Chico match factory. The abrupt decision affected a number of employees — 130 hourly production workers plus 20 management and office employees — and cost Chico a $1.3 million annual payroll. The announcement, which was made from New York, blamed the closing on a continued decline in sales of book matches, largely due to the greater use of disposable lighters. Kitchen matches, once a staple production of the factory, had gradually been phased out in the years following the Second World War, and the plant shifted entirely to the manufacture of book matches. Diamond will continue to produce matches at two other factories — Cloquet, Minnesota and Springfield, Massachusetts — but the corporation's decision to close the Chico operation ends 70 years of continuous match production in Chico.

The last log to be cut by the Stirling City sawmill, a large sugar pine, goes up the slipway on January 31, 1958.

The last log on the log deck before being rolled onto carriage at right. This marked the end of 54 years of operation.

The last log goes through the band saw. The off-bearer uses the hook to pull boards off the carriage.

Boards from the last log being fed into the edger by the edgermen. (All photos D. I. Corp collection)

The last crew of the Stirling City sawmill poses during their last day of work at the old mill. (Mrs. Mildred Forester collection)

Opposite: The stud mill at Stirling City. Built on the old mill site in 1964, it closed ten years later. (Western Ways, Inc., D. I. Corp. collection)

After the 1931 Stirling City fire destroyed the company office, Diamond remodeled a former warehouse as its office.

In the mid-1960's, a new office building was constructed to replace the old warehouse. (Both photos Kent Stephens)

The Red Bluff Integrated Forest Products plant shown here with identifying captions shortly after completion in 1959. (Gabriel Moulin Studios photo, D. I. Corp. collection)

The first carload shipment of 300,000 molded egg cartons left Diamond's Red Bluff plant on August 30, 1957, for Los Angeles. (Lloyd Maxson photo, D. I. Corp. collection)

A pond boat is used to tow logs on the 40-acre log pond at the Red Bluff plant. (Lloyd Maxson photo, D. I. Corp. collection)

An entire truckload of logs is lifted off as a unit by the unloader at the Red Bluff plant. The logs then go to a sorter, are strapped in bundles by species and dumped into the pond or cold decked for storage. (Lloyd Maxson photo, D. I. Corp. collection)

Aerial view of the Chico plant – The large U-shaped building at top center is the match factory. (Western Ways, Inc., photo, D. I. Corp. collection)

Combs of book matches being produced on a set and dip machine in the Chico match factory with Mrs. Marie Carlos in background. The large roll of splint board (called paper) on the right is being fed into the machine. The combs are cut from the paper rolls, then go through a pan to be dipped on their striking end. They continue to a booking machine where they are cut apart and printed covers stapled to them. (D. I. Corp. collection)

S. A. W. (strike anywhere) kitchen matches being boxed by automatic packer in the Chico factory in 1957. Circular row of boxes in lower center are on a revolving conveyor. Matches are sorted in the chute and dropped into boxes, fifty at a time. Then fifty more are dropped into the box, turned the opposite way. The Chico factory had eight of these machines; now there are none, as production is limited to book matches. (D. I. Corp. collection)

CHAPTER EIGHT

Devils and Angels

MANY PEOPLE believe that a logging town was a rough one in its earlier days. That was not true of Stirling City. In many respects it was similar to Lyonsville. Both had their settled, stable group of families who had their social life. Stirling had several social organizations, including the Odd Fellows, the Freemasons, the Eagles, and the Modern Woodsmen of America. Women's organizations included the Rebekahs, the Ladies Community Circle, and the Ladies Afternoon Whist Club. Couples often met at each other's homes for an evening and groups would spend Saturday night dancing at the hall in the merchantile building.

There was still another side to Stirling City during its early days. More than 600 men worked in the area, and prior to World War I families were not the rule in lumbering operations so much as at a later date. Most frontier towns had two classes of people: Stirling City was no exception. The two elements did not mingle in the moral aspect and a "decent" woman could walk down the sidewalk with Stirling City full of loggers and millhands on a pay day bender — and be safe. This is not to say that Stirling City wasn't a lively town or that the "fun" couldn't get rough.

Several forms of business activity catered to the various needs of the numerous single men: liquor, gambling, and prostitution. The articles of incorporation of the Stirling Mercantile Company granted it the right to deal in all types of liquors. The original plans for Fred M. Clough and Diamond provided a means to control the use of liquor in the company-owned town by a simple expedient. This was a clause contained in the deeds to Stirling City lots:

> Provided, however, that this land is sold and conveyed upon the condition that the second party, his heirs and assigns, shall not use or cause to be used or allow to be used, directly or indirectly, said land for or any part thereof for the purpose of making or selling intoxicating liquors thereon, and should said conditions be broken, the title herein granted by this deed shall cease, revert to, and be vested in the said Fred M. Clough, his heirs, successors and assigns.[1]

But one lot on the northwest corner of Quartz and Manzanita didn't have the liquor prohibition clause in its deed. The Stirling Mercantile Company bought it and the only saloon in town was built there during late 1903.

The saloon was a one-story building with a roofed porch on both streets. Its exterior was painted red, possibly with some paint left over from the sawmill. Before long, the saloon, named "The Red Devil," became famed in the history of Stirling City. The origin of this name is unknown, although it could have taken its name from other lumber town saloons whose name contained the word "red."*

As the only saloon in town, The Red Devil proved a lucrative business. There was no close competition, since the land for miles around was owned by Diamond, which did not allow establishment of any saloon on the property.

A related business stood next door to The Red

*At least three such saloons are known. There was the Red Light Saloon in Muskegon, Michigan. Perhaps more famous was the Red Keg Saloon at Averill, Michigan. It was the scene of the legendary fight in lumberjack lore between Joe Fournier and Silver Jack, reputed to be the two mightiest lumberjacks in history. The story of this fight was set down in print by Stewart H. Holbrook in his book *Holy Old Mackinaw*. Closer to Stirling City was another saloon with the word "red" in its name. This was the Red Cloud Saloon, located outside the city limits of McCloud, California. The saloon was similar to The Red Devil in that it was the only saloon in a company-owned town.

Devil. This was a notorious house of prostitution called "The White Angel." It was owned and operated by a woman whom we shall call by the fictitious name of Alice in deference to possible descendants. The White Angel had no connection with the Stirling Mercantile Company or its saloon. Alice, however, could not have operated her house without the consent of Diamond's officials, but it is unknown if any of these gentlemen were in partnership with her. Both The Red Devil and The White Angel were accepted as a normal part of life and a necessary part of society in that era. Both were mentioned by name in the newspapers. By comparison, Chico at same time had twenty-three saloons and a red light district.

The White Angel boasted at least three or four prostitutes "working" throughout the month. This number was insufficient to supply the demand on pay day, so at least a half dozen more girls from Sacramento and other cities arrived in town for that occasion. Prostitution was not confined completely to The White Angel, as some of the girls arriving for pay day would go together to rent an empty house for a few days. After the pay day spree was over, the girls would leave until the next pay day. The going rate charged by the prostitutes was $2.50, which was almost a day's wages earned by a laborer for working ten or twelve hours.

Chico was well supplied with saloons in the early days. The count was twenty-three shortly after 1900. Their interiors were ornate as seen in these two views. (Both John Nopel collection)

All accounts credit Alice with operating her business in a reputable and orderly manner (at least, so far as such a pursuit could be reputable and orderly). Her girls were required to practice the highest health and sanitary standards of their "profession."

She also owned a pool hall, which was a sideline to the more profitable White Angel. It was on Laurel Street several doors from the mercantile building. The pool hall was where Alice first met Bill (also a fictitious name for the sake of descendants). Soon after meeting, they were married. He apparently never worked while at Stirling City.

Conveniently situated near both The Red Devil and The White Angel was the Stirling City jail. It was small, measuring twenty by twenty-four feet, and similar to the community's other structures in that it was frame. Two-by-six planks were laid on their widest side and twenty kegs of spikes driven so close together through them that anyone attempting to cut through the walls would run into them.

The lawman at Stirling City during the early years was Constable Clay Buchanan. He often had his hands full stopping fist fights when the "timber beasts" (loggers) and "sawdust savages" (sawmill hands) mixed it up during pay day sprees. Many of these fights took place at The Red Devil and at Alice's pool hall. One of Stirling City's bloodiest battles happened at the pool hall when Jacob Greer and another man tangled, using knives and cue sticks for weapons. Greer, an old-timer in the area, had faced execution twice for killing a man in 1861 at Butte Mills.*

Diamond faced a challenge in 1906 to its liquor monopoly. Jack P. West, operator of a Stirling City grocery store, filed a petition for a liquor license with the Butte County Board of Supervisors. Both his first application and a subsequent one were defeated through the efforts of Major A. F. Jones, Diamond's

*Jake Greer was accused of the murder of Dutch Fred at Butte Mills (popularly called "Dogtown" and later renamed Magalia) on Election Day, September 4, 1861. The argument apparently erupted after Jake joined a group cheering for the Confederate States President Jefferson Davis. Dutch Fred objected to the cheering (Jake was later to say it was the opposite way). Jake insisted Dutch Fred join him for a drink: the latter refused and before long picked up a pick handle at a nearby store. Jake's weapon was a butcher knife: during their fight he stabbed Dutch Fred below the heart, a fatal wound that he died from six hours later.

Jake was arrested and was tried in Oroville during December, resulting in a hung jury. He was tried again, the second jury convicting him of second degree murder and sentencing him to be executed.

As Jake watched a scaffold being built, his execution was stayed by the California Supreme Court. Then in April 1863 the same court affirmed his sentence. Again the gallows was erected but an hour and a half before the execution, a document arrived from Governor Leland Stanford commuting his sentence to life imprisonment. He spent a decade in San Quentin before being pardoned.

An interesting sidelight to old Jake Greer is a story related by Anne E. Warren, a longtime resident of Magalia, in a letter to Chico State College folklorist Dr. Hector Lee in 1957. The letter was published in the Butte County Historical Society *Diggin's* as part of an article on Jake, and is used with permission:

". . . . I was working as a waitress in Chas. Bader's Hotel in Magalia about 60 years ago when Jake returned to Magalia one Sunday afternoon. The saloon was a room that joined the dining room. The men were drinking and somewhat the worse for liquor that afternoon. Just as we were closing the dining room, Bader brought in a very unprepossessing elderly man, told me he was Jake Greer, an Old Timer in Magalia and asked me to serve him supper.

"Greer asked for a steak 'well done.' I either forgot the 'well done' part of the order or the Chinese cook was angry at having to cook a supper when he had the kitchen ready for the night, so the steak was very rare when I brought it in. Old Jake roared what he would do to that Chinese So and So. Said he had stabbed a man in '61 for less than that. He would carve that Heathen's liver in little pieces. I grabbed the plate and rushed it to the kitchen where I found the cook had left. I slapped the steak on the top of the still hot stove, put a heavy hot flat iron on it, and it was soon well done.

"A very scared 18 year old girl had to stay in that big lonesome dining room while that partly drunk man roared on and gave me all the details of that murder and how he had to stay in prison in Oroville and watch the carpenters twice build and once destroy the gallows that was to hang him on. How he learned the harness makers' trade in San Quentin and when paroled was sent to Leland Stanford's ranch in the valley. How he worked there until the ranch was broken up and sold when he returned to Magalia.

"Greer became the roustabout at the hotel and spent several years there. He was a very quite [*sic*] peaceable man when sober. He was seldom drunk but whenever he was, he always recited all the details of the murder and his experience in the years he spent in prison.

"When Stirling City was founded in 1903 or '04, Old Jake went there as work was plentiful. He did general chores around town and in the summer he prospected in the hills.

"About 51 years ago he was helping Ralph Crumm move a small flimsy woodshed in Stirling City. The shed was on rollers when it started to collapse, and Greer, who was in back and outside, ran around and in the door just as it fell in. He thought that Crumm was inside. He had been but ran out and around the other end when the building started to fall. The roof came down and crushed Greer's chest.

"Old Jake always wore old-fashioned boots that required use of a bootjack to get off. He had a horror of 'dying with his boots on.' The only thing he said when my husband and several other men carried him out was, 'Boys, don't let me die with my boots on.' Someone took them off and Old Jake was dead. He is buried somewhere in one of the 'Unknown' graves in the beautiful little cemetery at Stirling City.

"Thinking you might be interested in the end of your story, I am

Respectfully yours,
Mrs. Annie E. Warren

". . . By the way when Old Jake used to tell the story of the murder he said the other fellow hollered 'Hurrah for Jeff Davis' when everyone was rejoicing over a Union victory. Jake was a German and was very fierce looking when he would draw back his lips from those old yellow snags of teeth and with his fierce beak of a nose. He would snarl like an animal over the murder. No wonder a very scared little girl never forgot the story!" [2]

Old Dogtown, today known as Magalia, at the time Old Jake had his election day fight. (Ralph Hupp collection, courtesy John Nopel)

Reprinted from Chico Record, *April 26, 1907.*

STIRLING CITY SAWYER AND OFFICERS CLASHED

ED SCOTT TRIED TO RUN TOWN, AND DEPUTY SHERIF PECK, BUT WAS LANDED IN JAIL.

Only Fight He Won During the Strenuous Evening Was in Beating His Wife, and He Was Thrashed by the Superintendent for Doing That—Lost Fine Position.

"Ed" Scott, one of the sawyers at the Stirling City mill of The Diamond Match Company, is down and out after having at sundry times ruled the uptown section when upon one of his "periodicals." Wednesday night he was full of "tea" of a bad brand, and undertook to "rough-house" the Hotel Raynor. Peaceable efforts to quiet him were unavailing.

Deputy Sheriff Peck chanced to be in Stirling at the time and, after Scott had disgraced himself and the community by his foul language, purposed to place him under arrest. Scott is a man of powerful physique, about six feet in height, and prides himself on his fistic abilities. Peck was unarmed and for a minute it appeared that it would be well for him to exercise the "better part of valor" and run, but he stood his ground, and Scott did likewise. Scott was armed with a quart bottle of whisky, which he flourished in a threatening manner, at the same time warning the officer not to lay hands on him. Finally the opportunity for Peck came, and he landed a straight drive to Scott's chin, putting him out for an instant and cutting a hole at the lower edge of his lip large enough to admit the passage of a lead pencil, the wound being inflicted by a ring setting on Peck's finger. Scott came up quickly and succeeded in breaking away from Peck, who did not pursue, believing that his arrest could be accomplished easier by two.

Scott ran for his home, and on the way hooked up with a closeline that came near causing him serious injuery.

Upon his arrival at home, he dragged his wife from the house, abusing her and using the foulest of language.

Superintendent Wisdom, who resides nearby, took a hand at this time, and gave Scott a beating that temporarily subdued him. Scott returned to his home and was engaged in washing the blood from himself when Peck and Constable Buchanan arrived. He met them at the dor with a stick of wood, and warned them that he would use the "Teddy Roosevelt" weapon on them if they entered. The "big stick" question was argued for some time, and to the end that for a moment Scott became friendly. He first shook hands through the partially opened door with Constable Buchanan and then with Peck. No sooner had Peck a firm grasp on his hand, than Scott found himself on the porch and Buchanan and Peck overpowering him. He put up a fight that was interesting for the officers, but they finally secured his hands with the steel cuffs without beating him.

The prisoner was ushered down to the town jail and locked up for the night. He has now lost his prestige as the boss of the town when full, and a fine position with The Diamond Match Company, he having been in charge of one of the band saws.

MISTOOK BUCHANAN FOR ONE OF THE ROBBERS

There was an amusing incident in the course of the journey of Deputy Sheriff Peck and Constable Clay Buchanan with Prisoner Matt Jackson from Oroville to Stirling City yesterday, and the joke was on Buchanan. As a precautionary measure, Buchanan handcuffed himself to the prisoner, Jackson, and they rode from Barber to Stirling in the baggage car.

At each of the stations along the line, there were groups of people gathered to see the prisoner, who with a pal had held up about sixteen loggers. They swarmed about the car at Paradise and Magalia, and at the latter place one man, who thought himself a deep student of human nature, entered the car and made a critical examination of countenances. He thought that there were two prisoners and that Deputy Sheriff Peck was in charge of them. After studying the countenances of both Jackson and Buchanan, he expressed the opinion that he thought Jackson "a pretty fair sort of a fellow," but he advised the officer to beware of that "sandy complected man with the faded moustache," for he certainly was a bad one. Buchanan didn't mind the compliment, but he is provoked to think that Conductor Willis heard it.

Reprinted from Chico Record, *August 25, 1908*

James A. "Teddy" Peck outside Fairburn Hall at the Barber plant. A Butte County Deputy Sheriff, he also served as Special Agent for Diamond and assisted Constable Buchanan in Stirling City. (Mrs. Alma Blasingame collection)

Pacific Slope attorney, who appeared before the Board. At the same time Fred Clough filed a suit against West, asking for return of the property and damages. Apparently it was settled without ever going to court and in April 1907, Jack West sold his store and headed for the Nevada mining country. It became all too plain to him that he could not carry on a fight with Diamond on two fronts—the company town and in the courts—at the same time. Very plain and simple, the law in Stirling City was The Diamond Match Company.

Less than a year later, the Stirling Mercantile Company's liquor monopoly ended. The force was the California Supreme Court decision against Galen Burdell, original owner of a tract around his Point Reyes Hotel in Marin County. Burdell's subdivision had deeds, similar to those at Stirling City, prohibiting liquor sale on the property, and violators suffered the same penalty. A court case started over this liquor restriction, and progressed through the lower courts to the California Supreme Court. Its decision on November 30, 1907, left the Stirling City authorities with little alternative:

> The court held that in a suburban district where lots were sold under restriction, the district must be either entirely dry or wet. There can be no exception even in favor of the owner of the tract.[3]

The court's decision gave Diamond the alternative of either allowing other saloons in Stirling City or making the town dry by closing The Red Devil. They chose the latter course of action and a representative of the Stirling Mercantile Company explained the decision to the press:

> The stockholders of the Stirling Merchantile Company are men connected with The Diamond Match Company. They have held it was better to have but one saloon and have it under their control, than to open up the town to saloons, for the reason that it would be almost impossible to keep employees of the mill and logging camps from frequent dissipation.
>
> Financially the Stirling saloon has been a winner, but the stockholders have not been pleased with it, for they have found that it is impossible for them to have their orders, relatively to the conduct of the place, obeyed. Consequently for some time past they have been considering the advisability of closing the notorious 'Red Devil' for good . . .[3]

The Red Devil closed in approximately March 1908, but by then the Stirling Mercantile Company had determined a method to retain the liquor monopoly and at the same time stay within the law. The company purchased several acres from Diamond in Stirling's suburb south of town that was known as "Dago Town." A new saloon, called "The Midway," was built on the land. Thus the company stayed within the law by going outside the townsite itself.

When The Red Devil closed, Alice found that her business slowed without her neighbor. Within a year, she closed The White Angel. Alice and her husband left Stirling City for Hamilton City, then booming from construction of a sugar factory. The same court decision that upset plans in Stirling City resulted in Hamilton City becoming a wide open town for all saloons that qualified for licenses. Their move to Hamilton City brought them nothing but trouble. Complaints against the saloons resulted in many of them being closed and it was difficult to obtain enough signatures for a new license application. Their saloon was closed and then attached for unpaid wholesale liquor bills. At that point Bill deserted his wife. She managed to pay the debts and returned to Stirling City, where she operated the pool hall.

During her absence a new sporting house opened next to the Midway Saloon. The new madam, known only as Phyllis, leased a house for her establishment.

Gambling provided one of the favorite pastimes for the lumberjacks in Stirling City. During the first decade after its founding, the community became famed in Northern California as a "wide-open" town. Professional gamblers lived in the town and others arrived for the big games held on each pay day. These professionals were usually the winners in the games as most of the workers were unable to match their skill at cards. Most of the games took place at Alice's poolhall.

The professionals met their match one time, however, and came out on the worst end. They were taken by another professional gambler who had lived in Stirling City for a few months, posing as a logger. He spent his spare time watching the games to detect the weaknesses of the professionals. Finally one night he went into action, playing stone sober, strictly as a business proposition. By the time the game broke up, he had nearly all of the money the professionals and their friends could gather and left town with between $2,000 and $4,000 to the good.

In 1910 a dispute among the gamblers led to so much notoriety that the "lid" was clamped on the gambling. One of the gamblers became involved in a game at the pool hall and was provided with free drinks to the extent that he became very drunk and finally fell into a stupor. He filed a complaint a few days later with Constable Buchanan, charging the dealer with robbing him of his missing $500. Buchanan arrested the dealer, but he remained free pending a preliminary examination.

During the next few days, Constable Buchanan, using this incident as a pretext, issued a warning that the professional gamblers and their friends were not welcome as residents. The gambling houses became virtually deserted within days as the gamblers took Buchanan's advice and left town in a general exodus. Buchanan then started closing down all of the gambling "resorts" as the next move in purging the town of the gambling element.

Among those leaving permanently at that time was Alice. Her dealer, forced to stay to face robbery charges, was virtually the only professional gambler left. He never really faced a trial, as the case was continued indefinitely because the complaining victim left town for an "unknown" destination.

Many years later Clay Buchanan said in an interview that it had not been the intention to completely close gambling for all time at Stirling City, but only long enough to get rid of the professionals. Gambling did end at Stirling City by law shortly afterward. On March 9, 1911, the County Supervisors passed an ordinance that banned gambling within all unincorporated areas and limited the opening hours of saloons.

Alice's pool hall, a sideline to her more lucrative White Angel, was the third building from the left. Just beyond was the large two-story Robe Hotel with the long porch in front. (John Nopel collection)

CHAPTER NINE

The Butte County Railroad

THE Butte County Railroad played a major part in Diamond's early efforts to promote Stirling City as a resort area. The company hoped to attract not only week-end vacationers, but also those who would purchase summer cabins and use the railroad to commute to work in the valley.

Frequent excursions provided the major part of the promotion. The first excursion had celebrated completion of the railroad to Magalia. The railroad received its new passenger equipment in May 1904, and later that month a "Back to the Woods Excursion" carried 150 people to Stirling City. An even larger excursion came on June 11, 1904, when Diamond gave free tickets to more than 1,000 people. Locomotive No. 4, a brand new 2-8-0 built by the famed Schenectady engine works, handled the nine car consist, which included open cars. The Chico Park Band, riding one open car, provided entertainment enroute.

Excursions to the mountain town proved popular, with any holiday or special event serving as excuse for an excursion sponsored by local organizations in Chico or Stirling City. Special facilities were built at Stirling City to care for the tourist traffic. These included two hotels, cabins, a baseball diamond, and an open-air dance pavilion. Trains carried revelers to the dances, arriving about 9:00 p.m. in the evening and leaving at 3:00 a.m. the following morning. There were snow excursions for a number of years during winters. The largest excursion ever handled by the railroad was on July 4, 1910, when two special trains took more than 3,000 people to Stirling City for the Independence Day celebration.

As part of the promotion, the railroad in 1904 issued a booklet called *Pine Chips,* with text and illustrations of the scenery along the railroad and at Stirling City.

After 1910, the excursions began to decline in number. The early dreams of making Stirling City a resort center failed to materialize and it remained primarily a mill town. The lack of population in Butte County and the transportation distance from larger centers of population hindered the town's development as a resort. When the program floundered, company interest in the promotion declined. Another factor was the resignation of Fred Clough, who had regarded Stirling City as his personal project and had taken great interest in its development.

The main purpose of the railroad, however, was to haul lumber from the Stirling City mill. Lumber, "hogged fuel," and slabs were not the only items on the conductors' waybills. Other tonnage included logs, livestock shipments, apples, ore, and general freight.

During the first two months the railroad was in operation, the schedule called for a daily round trip by a mixed train. It left Chico at 7:05 a.m., arriving in Stirling City at 9:45 a.m. The return trip left Stirling City at 3:15 p.m., with arrival in Chico at 6:00 p.m. From Stirling City, travelers could take stages for Big Meadows (Prattville) and other points in Plumas County.

A new schedule inaugurated June 30, 1904, replaced the single round trip of a mixed train (passenger and freight combined) with a daily passenger and a daily freight train. This schedule was usually in effect only during the summer while the sawmill operated. During the winter months, the schedule was cut back to a daily round trip by a mixed train. With some variation, this remained the pattern of the railroad's operations until its days as an independent company ended.

In common with other lines of the early 1900's, the Butte County Railroad had accidents, but they do not

No. 3 bucking deep snow at Stirling City on a passenger train run during the winter of 1907-08. (D. I. Corp. collection)

A passenger extra bound for Chico pulls into Magalia behind No. 4. This view was from the second-floor living quarters of the agent. (Anne Warren collection)

No. 4 decorated with bunting and flags for an Independence Day excursion to Stirling City. (Wayne Crowder collection, courtesy John Nopel)

seem to be an excessive number. For one in mountains, the Butte County Railroad was probably safer than some of its contemporaries. The company did have trouble on several occasions with the I.C.C. over safety appliance violations (grab irons, etc.).

Some accidents were tragic; others bordered on the semi-comic. On one occasion, a section man was coastting downgrade on his gravity car loaded with a number of carpenter tools. Suddenly a bar fell off the gravity car and derailed it, pitching its occupant off head first. He was lucky and returned to the job ten days later. Another strange accident involving the section men happened at quitting time on the afternoon of August 3, 1909, at the Barber section house. A gravity car, with a barrel of water with two men holding it, was standing on the track. The section crew came in on the same track aboard their hand car. As they approached the stationary gravity car, one of the section men shoved it with his foot. The car moved forward, but the barrel tipped over. In their efforts to save the barrel, the two men on the gravity car were pulled off and the barrel and both men landed on the track directly in front of the approaching section car. The men scrambled to safety but the section car struck the barrel with great force, injuring one man on it.

A semi-comic accident occurred on January 7, 1907, when 4-4-0 No. 3, plowing snow between Stirling City and Magalia at 30 miles-per-hour, rounded a curve to find a man walking down the middle of the track just ahead. There was no time to stop on the icy rails and the man was plowed off with the snow, landing in a drift and sinking almost out of sight. The crew stopped to dig him out, found him uninjured, and gave him a ride into Magalia in the engine cab.

The following summer another lucky "accident" victim was a young woman, who boarded the afternoon passenger train at Stirling City and took a seat in the rear car. Just as the train neared the Doon trestle, she rose from her seat, bowed and smiled at the woman sitting across the asle, and walked to the rear platform of the coach. The woman that had been sitting across from her feared something was wrong, and went to the rear platform in the company of a friend. They saw the young woman lying in a heap alongside the track within a few feet of the Doon trestle. One of the other passengers attempted to stop the train by pulling the cord and when that failed, went to tell the conductor. The train stopped and backed up to where she lay. She was placed aboard and taken to Chico, where she was confined to a hospital. Her attempted suicide, apparently over marital troubles, resulted in only a few bruises. She was conscious during the ride to Chico and explained her

American Standards Nos. 2 and 3 doublehead an upbound train in Magalia. (John Nopel collection)

	ALTITUDE FEET	STATIONS	DISTANCE FROM BARBER	
Principal Points on the Line	193	CHICO	1.	of the Butte County R'y
	186	1 MILE BARBER	0	
	....	3. BUTTE CREEK	3.	
	825	4.5 CROUCH	7.5	
	1708	5.7 PARADISE	13 2	
	2321	4.9 MAGALIA	18.1	
	2700	3 PARKERS	21.1	
	2802	4 DOONS	25.1	
	3525	5.5 STIRLING CITY	30.5	

Excursion train at Stirling City about 1911. George Bell, engineer; Fred Haskins, fireman; Bill King, conductor; and Clifford Whiteside, brakeman. (Mrs. Fred Haskins)

Butte County Railroad's "big power," 2-8-0's Nos. 1 and 4, in front of the Barber enginehouse about 1906. This was Barber' first permanent building. (D. I. Corp. collection)

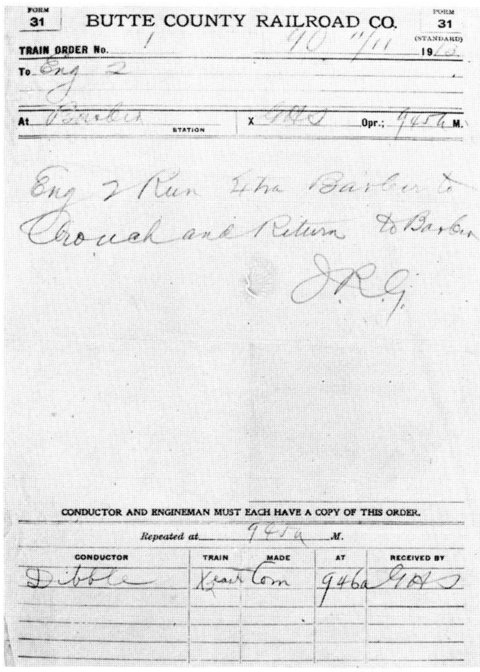

FORM 31 BUTTE COUNTY RAILROAD CO. FORM 31 (STANDARD)

TRAIN ORDER No. 1 19__

To Eng 2

At Barber STATION X Opr.; 945a M.

Eng 2 Run Xtra Barber to Crouch and Return to Barber

JRG

CONDUCTOR AND ENGINEMAN MUST EACH HAVE A COPY OF THIS ORDER.

Repeated at 945a M.

CONDUCTOR	TRAIN	MADE	AT	RECEIVED BY
Dibble	Xtra	Com	946a	

Conductor C. R. Dibble saved this memento of his service with the railroad. "JRG" was John R. Gray, then railroad superintendent.

A lumber train ready to depart Stirling City behind No. 1. (California State University, Chico Library collection)

actions by saying "she wanted off but didn't have the time to tell the conductor."

Not so fortunate was an Italian section hand employed by the railroad, who had his accident in the same place a few years later. His undoing was choosing the center of the track just above the Doon Tunnel for a nap. A Chico-bound train struck and killed him before the engineer could stop. The section hand's body was brought to Chico aboard the same train that had killed him.

It was on June 25, 1909, that the Butte County Railroad had the worst wreck in the history of its operations. The schedule called for the upbound passenger train to leave Barber at 4:05 p.m. and to arrive at Paradise at 4:43 p.m. The freight for Barber was scheduled to leave Stirling City at 2:00 p.m. with arrival at Paradise at 3:43 p.m. Usually the two trains passed each other at Crouch Siding, 5.7 miles below Paradise and 7.15 miles from Chico. On this day the freight was one hour and twenty minutes late, leaving Stirling City at 3:20 p.m. and Magalia at 4:38 p.m. The passenger train was also running late, and didn't leave Barber until 4:12 p.m. It was due in Paradise at 5:15 p.m. with the freight scheduled to arrive at 4:56 p.m. The accounts in Chico's two newspapers vary on some of the details of this afternoon, and the schedule was just one of the differences. According to the *Chico Record* the upbound passenger train was hauling an extra load of seven freight cars, which slowed it on the grade sufficiently that it couldn't make Paradise in time for the meet with the freight. The rival newspaper, the *Chico Enterprise,* said the passenger train was ordered to take siding below Paradise and the freight was given a clear track to the

Piles of wreckage still smoldered the morning after the wild runaway of the lumber train at Paradise in 1909 (above). At left, the salvage crew attempts to move two tanks covering the unfortunate brakeman Amos. Below, the section gang can be seen rebuilding several hundred yards of track torn up by the wreck. (Top, Mrs. Justa Farley collection; others, Mrs. Lettie Lillybridge collection)

siding.

The passenger crew consisted of Conductor A. J. Willis, Brakeman W. L. King, Engineer George Bell, and Fireman Fred Haskins. An extra crew was handling the freight due to the regular crew working overtime the day before, when they had to clear the line after the rear axle on the freight engine's tender sheared and derailed a box car. Conductor Albert Johnson (spelled Johnston in most of the newspaper stories) was in charge of the freight train, and its crew, Head Brakeman Earl D. Amos, and brakeman J. H. Hume, and William L. Lewis. J. F. Myers was firing the engine and John B. Nesbit was engineer.

The railroad regularly carried fare paying passengers on the freight and six passengers were aboard the caboose that day. All went well until after the downgrade, 27-car freight passed Magalia. Through Magalia there is a "sag" in the otherwise steep grade, where the track is fairly level before the steep grade resumes. At Optimo the grade is 2.7 per cent increasing to 3.4 per cent at Paradise.

At this point the passengers received the scare of their lives and one of them related the experience to a reporter afterwards:

> "We had passed Magalia and everything was seemingly all right. There were five other passengers in the caboose besides myself. I was seated and dozing, when suddenly I came to the realization that the train was going at a high rate of speed. I went to the rear door and looked out. I then realized something had happened. Trees and landscape went by like lightning, and the roar of the train was deafening. Great clouds of dust swirled under the caboose and were sucked back into the car. I then crawled up into the lookout on the caboose and saw that we were running away. The brakemen along the tops of the cars were twisting the brakes with all their might. Conductor Johnston ran from car to car, doing all that he could. Through the dust I could see the forms of the brakemen holding onto the brakes. By this time the rest of the passengers had become aware of what was happening. An old man who was in our car insisted on jumping from the back platform, but the rest of us held him inside. The car jumped and bumped from side to side, and I thought that we would leave the track at any moment.
>
> "Conductor Johnston then came into the car. He was cool and said: 'We are running away, and I will do the best to save your lives.' He then went to the front platform of the caboose and, holding onto the iron rail, reached down and drew the coupling pin. He already had the brakes on full strength, and we redoubled our efforts to stop the car. It was an exciting moment for us. We did not know whether the speed of the car could be lessened or if it would follow with the rest of the train. Soon, however, there came an inch of space between the caboose and the next car. We tugged and hauled at the brakes, and finally the rest of the train gradually pulled away from us and the air coupling snapped. We were still going, however, at a high rate of speed. The train thundered on, winding in and about the curves. By the time we reached Paradise we had slowed down and stopped there. Conductor Johnston deserves great credit, for had he not acted as he did we would all have been killed or injured. There was one woman in the caboose and she acted bravely all through the ordeal."[1]

Despite the action of the handbrakes, the runaway roared through Paradise at a speed of more than a mile a minute. (The downgrade speed limit for freights was only 16 to 18 miles-per-hour.) As the engine tore down on the Paradise station, Agent Miller saw Nesbit tugging on the jammed reverse lever. About 100 feet below the depot, Nesbit "joined the birds" jumping from his runaway engine to land head first on a pile of rocks and gravel. The remainder of the crew still aboard undoubtedly knew by this time that the train would never make it around the curves below Paradise.

The train rounded two curves and approached the third, three-fourths of a mile below Paradise at Neal's Crossing. It bent to the right in a six degree curve atop a fifteen-foot fill on a three per cent grade. The runaway freight roared into the curve at a speed of nearly 70 miles-per-hour. Part way around the curve the car just behind the engine left the rails. The oil tank on top of the tender frame pulled loose and followed the first car over the right side of the embankment. Within seconds the cars behind the first carload of lumber left the rails and with a deafening roar, heard for miles, piled on top of the first carload of lumber and the tank portion of the engine's tender. Car after car went to pieces and in a few seconds a mountain of wreckage — cars, lumber, logs, and slabs — filled the ditch below the embankment to a depth of fifteen feet or more for a length of 100 feet.

The three brakeman still clung on the cars when the runaway tore into the curve. The nearest one to the engine was Amos, who was two cars back. As the train rounded the curve, he climbed down off the logs on his car and then jumped from the train on the right side just before the cars started piling up. He was buried under the wreckage. The other two brakemen were more fortunate. Lewis was on a carload of lumber seventeen cars from the caboose when his car went into the ditch. He was caught and pinned under lumber. Brakeman Hume was ten cars from the caboose and seven cars from Lewis. He managed to stand on top of the carload of lumber until it went to pieces under him, and found himself unhurt on top of the lumber in the ditch. As he picked himself up he dis-

EASTWARD			Time Table No. 156 June 13, 1937	WESTWARD	
FIRST CLASS					SECOND CLASS
Capacity of Sidings and Spurs in Car Lengths	526 Mixed	Distance from San Francisco	Stirling City Branch	Distance from Stirling City	527 Mixed
	Leave Mon., Friday		STATIONS		Arrive Tues., Saturday
Yard OYWBKP	1.15 PM	184.2	TO-R CHICO	31.2	s 10.40 AM
			2.4		
		186.6	S. N. R. R. Crossing	28.8	
			2.7		
	f	189.3	DREDGE	26.1	f 10.20
			3.3		
18	f 1.50	192.6	CROUCH	22.8	s 10.00
			5.6		
21 WP	s 2.20	198.2	PARADISE	17.2	s 9.30
			5.0		
12	s 2.50	203.2	MAGALIA	12.2	s 9.05
			3.0		
3 P	s 3.10	206.2	APPLETON	9.2	s 8.50
			3.8		
10 (Spur) WP	s 3.30	210.0	DOON	5.4	s 8.30
			5.4		
PY Yard	s 4.00 PM	215.4	R STIRLING CITY	0.0	8.00 AM
	Arrive Mon., Friday		(31.2)		Leave Tues., Saturday
	(2.45) 11.35		Time over District Average Speed per Hour		(2.40) 11.70

Additional Flag Stops to Receive or Discharge Passengers: Nos. 526, 527, Barber (Chico yard). Oakdale Farms, Ostrander, Wagstaff, Optimo, Bush and Luce.
Additional Stations: Barber 185.1. Oakdale Farms 196.6, Ostrander 199.6, Wagstaff 200.2, Optimo 202.2, Bush 203.9, Luce 207.1.
Emergency water supply, Stirling City

Southern Pacific Sacramento Division employee's timetable, Stirling City Branch. (Bill Shippen collection)

covered he was only twenty feet from Lewis, who was yelling for help where he was pinned under the lumber. Hume scrambled over the jumbled mass of lumber to Lewis and extricated him without much difficulty.

The newspaper accounts varied on Fireman Myers. The story in the *Chico Record* claimed he leaped about 100 feet before the crash, but the *Chico Enterprise* stated he had an exciting ride until the engine came to a stop. Apparently the tender fell apart all at once, the oil and water tank above the frame going over the edge of the embankment with the first car, with the trucks derailing into the ditch. About the same time all four drive wheels dropped onto the ties, but the front pony wheels of the 2-8-0 remained on the rails. This was probably the only reason the engine stayed upright on the track as its momentum carried it 200 yards down the track beyond the wreck. This last 200 yards of Fireman Myer's ride must have been the most terrifying of his chilling experience. With the tender gone, he had only the thin iron apron to stand on as he hung on in the cab. As the drive wheels bounced over the ties, Myers was hurled back and forth in the cab, suffering a dislocated ankle and wrist as well as very severe body bruises.* Had Myers fallen backward out of the cab, certain death would have resulted since one part of the tender remained attached to the engine. This was the frame, secured to the engine by a heavy chain that held after the drawbar snapped. The frame, bouncing along behind the jolting, rocking engine, would have killed Myers had he landed on the track behind the cab. The condition of the engine impressed a reporter for the *Chico Enterprise,* who wrote:

> Approaching the engine from the front appearance gives no indication of what the engine went through, but from the rear and sides a different sight is presented. The bare frame of the platform on which was mounted the oil tank lies crosswise of the track, attached to the engine by a heavy chain . . .
>
> Along the track for the distance between the engine and the wreck lie scattered pieces of machinery of the engine, and the ties are broken and the rails torn up in a manner which causes no end of wonder of how Myers succeeded in remaining at his post.[2]

Only three cars at the end of the train stayed on the track. Within minutes after wreck, the pile of lumber and wrecked cars caught on fire from a hotbox that was to have been fixed when the train stopped at Paradise.

The two newspapers varied again in their accounts of the passenger train. The *Chico Record* reported that Brakeman Hume hobbled down the track to flag the upbound passenger train, which arrived at the wreck five minutes later. According to their account, the delay of adding the carloads of machinery to the train at Barber plus the additional delay of the extra load on the grade saved the passenger train, which otherwise would have met the runaway freight in a terrific head-on collision had it been on time. The reporter for the *Chico Enterprise* wrote that a head-on collision was impossible, due to the passenger train taking siding below the point where the freight piled up.

Just about the time the freight piled up, Chief Clerk S. B. Nourse at Barber attempted to call the Paradise agent, but found that neither the telephone nor the telegraph lines were working. The wreckage was directly below the wires, which were either broken by the wreck or destroyed in the subsequent fire. A few minutes after Nourse had tried to call, Conductor Willis of the passenger train hooked up an emergency phone (carried on all Butte County trains) from the baggage car and notified Nourse of the disaster.

Nourse promptly obtained an yard engine from the Barber yard to power an emergency relief train to the scene of the wreck. Aboard it were Dr. M. P. Stansbury (Diamond's local physician), two nurses from the Sisters' Hospital, Butte County Railroad General

*Mr. Myers was crippled for the rest of his life from injuries received in the wreck.

Superintendent John R. Gray, Special Officer J. A. Peck, Master Mechanic James A. Chambers, Carshop Foreman W. E. Laramore, and other officials. Two relief trains were dispatched from Stirling City with every available man aboard under the supervision of Roadmaster Pat Lyons. After the passengers, mail and express were conveyed around the wreck in wagons to one of the relief trains from Stirling City, the passengers from the caboose and injured were taken by the passenger train to Barber. One of the two relief trains from Stirling City was held at Paradise, pending work on reopening the line.

A number of Paradise residents had joined in the rescue work before the relief trains arrived. Frantic efforts were made to find Brakeman Amos under the huge pile of lumber and wreckage before the fire gained too much headway. The search continued for hours, with workers getting as close as possible to the fire. Many also formed a bucket brigade, which was effective in containing the fire to the wreckage. Others joined with Diamond's employees from Stirling City in saving more than 130,000 board feet of lumber from the flames.

No estimate of damage was released by Diamond, but one of the local newspapers stated that it was believed to total several hundred thousand dollars. The wrecked freight had consisted of twenty cars of lumber, three cars of "hogged" fuel (sawdust and chips for the Barber powerhouse), one car each of slabs and shingles, two carloads of logs for the veneer factory, and the caboose. All but the last three cars were derailed and wrecked. The train carried 308,000 board feet of lumber, 18,000 feet of logs, 140,000 feet of shingles, and an unknown amount of hogged fuel. The wreck tore up 800 feet of track.

Shortly before 3:00 the following afternoon the charred remains of Brakeman Amos was found in the ashes of the wreckage. It was thought that his head had fallen in the angle formed by two metal tanks and that he had been killed instantly under the carload of shingles. Engineer Nesbit was operated on at the Sisters' Hospital in Chico, but died around 2:00 a.m. Monday morning, June 28, without ever regaining consciousness.

A coroner's jury held two separate inquests on the train wreck on the afternoon of June 28. Their verdicts held that both men had died as a result of Nesbit losing control of his train through negligence. All witnesses agreed that the engineer had misjudged the grade and lost control of the train by letting it run too

The Paradise depot was the smallest of Butte County Railroad's four agency buildings. It is still standing today, little changed from its 1916 appearance. (Southern Pacific)

The Stirling City depot in 1916, just after the Southern Pacific assumed operation of the line. Evidently the S. P. sign painter didn't know how to spell the town's name – a common mistake. (Southern Pacific)

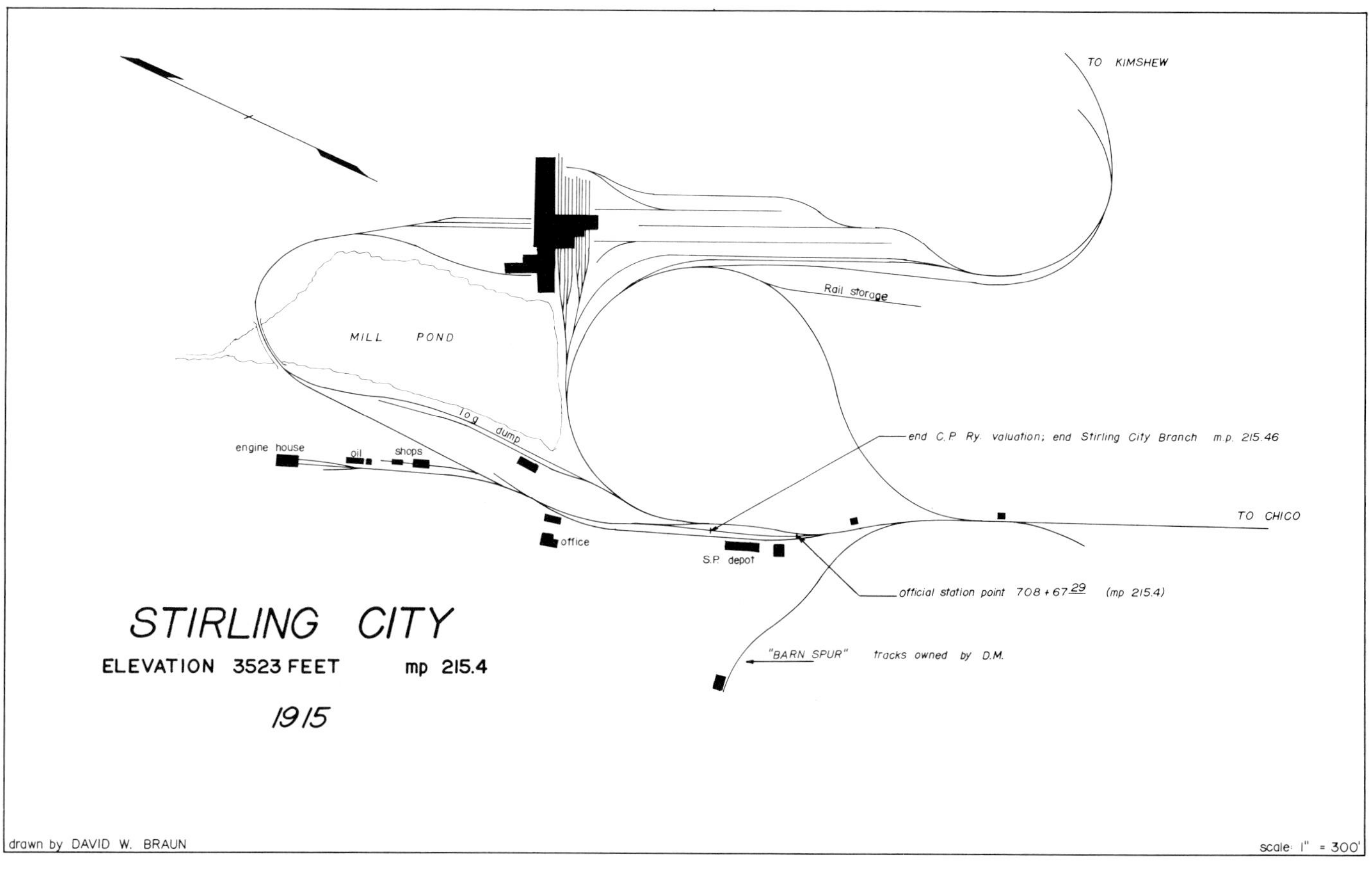

fast approximately one and a half miles above Paradise. He had full control of the train until after he passed Magalia, but as one of the other engineers said at the inquest, "he didn't catch his air quick enough" as the train left the nearly level track of "Magalia sag" and nosed over into the two per cent grade above Optimo.

Nesbit had only been hired a short time before the fatal wreck, but he had seven or eight years experience on other California railroads. The fatal trip was the first that he and the conductor had made on the railroad in charge of a train. He had made his student trip the previous day under another Butte County Railroad engineer, who stated Nesbit had done well except for a switching move at Stirling City. Another engineer testified that it was easier to handle a 27 car train than 16 cars on the grade due to the greater braking efficiency and more wheel friction on the curves. Conductor Johnson stated in his testimony that the release of the caboose didn't interfere with the engineer's operation of the air brakes. Master Mechanic Chambers, who hired engineers and fireman, said Nesbit's credentials were good and that he was careful in his operation of a train. All equipment was in good condition when the train reached Magalia except for the hotbox in a bar that was to have been repaired at Paradise. Trainmen at each terminal inspected all equipment and made necessary repairs. An examination of the wrecked engine and the positions of the levers and brake valves in the cab showed that Nesbit did everything he could to stop the train.

On February 29, 1912, the Chico & Northern was dissolved and its assets transferred through the Central Pacific to the Southern Pacific. This change in the lease did not affect the operation of the Butte County Railroad. It should be pointed out that the Chico & Northern was never an operating company and no equipment was lettered for it. Its sole asset was the lease of the Butte County Railroad.

Two years later it was the Butte County Railroad's turn to follow the Chico & Northern into limbo. In 1914 Diamond's Board of Directors faced the question whether to renew the lease or not. They decided to terminate it and turn the railroad over to the Southern Pacific on December 1, 1915. The possible reason for their action was the railroad had operated at a loss during 1914 and continued to do so during 1915. The difference between profit and loss was caused by the annual rental charge. The decision was part of the general cut-back of all of the company's operations in California during 1915 and 1916.

Superintendent Gray issued the last Butte County Railroad timetable (No. 38) on November 7, 1915 and trains operated under that schedule during the remainder of the month.

Preparations for ending operations of the Butte County Railroad began in late November. J. H. Dyer, Southern Pacific's Sacramento division superintendent, arrived in Chico aboard his business car on Sunday evening, November 28. The next morning a special train carrying Dyer and other S. P. officials, along with B.C.R.R. Superintendent Gray, made an inspection trip over the railroad. The final run of the Butte County Railroad came the following day, on the afternoon of November 30, when the regularly scheduled mixed left Stirling City at 3:00 p.m. with George Schmidt at the throttle and A. J. "Acey" Willis serving as conductor. Interestingly, it had been Willis who had fired the engine of the first passenger run on the Butte County Railroad when the excursion to Magalia

The rear of the Stirling City depot presented a less-interesting appearance. (Southern Pacific)

was made in 1903.

When the train tied up in Barber that evening, both the railroad and the crew were through. The Southern Pacific earlier notified the crews that their seniority rules would apply at the end of the lease and the Butte County men would be treated as new employees. The only B.C.R.R. men actually retained were the section hands and three trainmen on the Chico switch engine, which was put on regularly for the first time. Superintendent Gray was shifted to the position of traffic manager for Diamond and ultimately rose to be Manager of the company's Northwest Timber Division. After the retirement of W. B. Dean as Manager of the California operations in 1937, Gray was placed in charge of all Pacific Coast operations of the company. Vernor S. Woolley was General Manager of the Butte County Railroad when it ceased operations. Woolley inherited this title from Fred M. Clough when the latter resigned.

The Southern Pacific inaugurated a new schedule when it assumed operations on December 1. The morning train continued to leave Chico at 7:30 a.m. and arrive at Stirling City at 10:45 a.m., but the return trip was set forward three and a half hours with the train leaving Stirling at 11:45 a.m. to arrive back in Chico at 2:40 p.m. This permitted connection with both the "southbound motor" and No. 22, "The Statesman," running northbound. Both trains left Chico at 2:55 p.m.

The Barber depot was closed on the last day of the Butte County Railroad's operations and its business transferred to the Chico depot. Near the end of March 1916, a Southern Pacific crew started moving the Barber depot to the new terminal under construction at Gerber. The station was moved in sections to the new location. It was not until June 1, however, that the agency at Barber was discontinued by a permit issued by the State Railroad Commission.

The brick enginehouse built by the Butte County Railroad at Barber continued in use, as the Southern Pacific didn't have any engine facilities in Chico. If an S.P. engine needed servicing at Stirling City, it was handled at the Diamond Match Company's enginehouse. This seldom occurred.

All four of the Butte County Railroad's engines and most of the equipment were taken over by the new owner. No. 3, apparently out of service for some time, was towed to Sacramento in partially stripped condition and scrapped a few months later without being as-

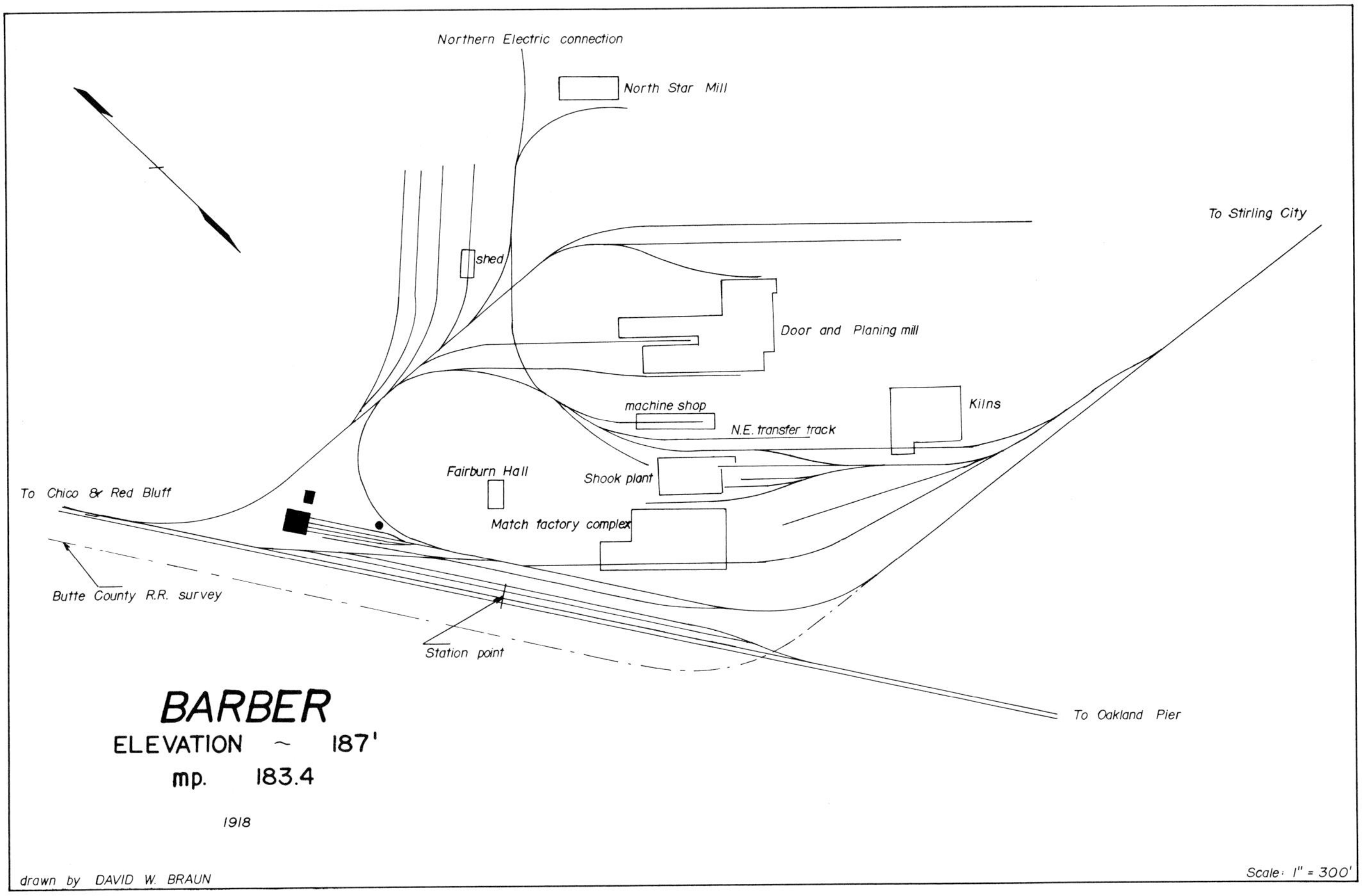

signed a new number. The same fate befell the other American Standard, No. 2, which also was scrapped at Sacramento three months afterward without ever being renumbered [S.P. (2nd) No. 1304]. The two Consolidations fared better, both were renumbered into the S.P. roster and saw active service until the early 1930s.

The Southern Pacific kept some of the Butte County's equipment right on its home rails. Several retired employees recall that the S.P. assigned 2-8-0 No. 2502 (2nd) to power the Stirling City mixed right after it assumed operation of the line. It is not likely that very many recognized the little Consolidation in its new livery as former B.C.R.R. No. 4. Keeping the one-time Butte County engine company on the mixed was one of B.C.R.R.'s combos, which the latter company had also used on the mixed run. No. 2502 remained the regularly assigned engine on the branch for a number of years with other S.P. locomotives assigned as needed for relief or extra trains. Largest power used were the small 2500 series 2-8-0's and the 2-6-0's numbered in the 1600, 1700, and 1800 series.

A fairly consistent pattern of operations characterized the Stirling City Branch for many years. The mixed train operated on a daily basis between Chico and Stirling City. Usually there was also a freight train on the branch during the periods that the Stirling City sawmill was in operations during the summer months. Every other day a round trip was made between Gerber and Stirling City. The small engines used on the mixed to Stirling City and the local freight were stabled in the old Butte County enginehouse. A flanger and a fire train water car were also kept there.

Most of the time the mixed was powered by No. 2502. Some time during the late 1920s or early 1930s, she was removed from the run. According to retired

The Magalia depot in 1916. Note the photographer's hand-powered velocipede in left corner. It must have been hard pumping up the 3 per cent grade! (Southern Pacific)

trainmen, the little Consolidation finally wore out. She was laid-up in Sacramento and was scrapped in 1935, preceding No. 2503 (former Butte County No. 1) by about a year to the scrap pile. Replacing No. 2502 on the Stirling City was 2-6-0 No. 1780, which stayed in service on the branch on a fairly regular basis for more than fifteen years.

In 1939 the Southern Pacific petitioned the California Railroad Commission for authority to discontinue mixed train service (Trains 526-527) on the Stirling City Branch. The Commission set hearings for February 15, and the railroad was given permission later in the spring to discontinue the train as a mixed. The service by that time had dropped to twice weekly.

Even before the mixed was discontinued, fire destroyed the Stirling City depot during the night of March 13, and for a time endangered the entire town and the sawmill. A new fire siren had been purchased through a public subscription by Stirling City's residents the previous year. It was given the credit for

The Barber depot was at the end of Del Norte Street (now Chico's West 16th Street) between Diamond's plant and the S. P. line. (D. I. Corp. collection)

Stirling City was served by S.P. mixed trains Nos. 526-527 for years until 1939. No. 1780 shown ready to leave Chico in 1937. (W. C. Whittaker)

The end of mixed service didn't affect No. 1780. The little mogul continued in service for years until replaced by larger 2-8-0's. Here she is shown, equipped with snowplow, at Diamond's plant in 1946. (Guy Dunscomb)

saving the lives of the agent and his wife, who barely had time to flee in their night clothes from their apartment in the blazing depot. The depot was not rebuilt.

In the winter of 1949-1950, the Doon Tunnel caught on fire and partially collapsed. It was sealed off and replaced with a by-pass line bulldozed around the lava dike. Major work on rebuilding the track started in 1946 and by 1956 the entire branch had been laid with used 110-pound rail on new ballast. After several of the bridges were strengthened, larger steam power was allowed on the branch. These included 2-8-0's of the 2700 and 2800 classes, which were the largest engines used on the line.

On June 10, 1951, a passenger train again ran over the Stirling City Branch. It was a railfan excursion special, "Weekend in Paradise," which had left the Bay Area on June 9th using regularly scheduled trains to Chico. The Chico Chamber of Commerce made overnight hotel arrangements for the excursionists. The following morning they boarded a special train, powered by 2-8-0, No. 2728, bound for Stirling City. Arrangements had been made with the Diamond Match Company for all Shays to be available in the Stirling City yard in good locations for photographs. Several photo stops were made during trip over the Stirling City Branch, both on the way up and during the return to Chico late in the afternoon.

By 1952 the Southern Pacific had diesels in regular operation on the branch and need for the old Butte County Railroad enginehouse for maintaining Chico-based steam engines vanished. Most of the steam engines were sold to various scrap dealers or cut-up by the Southern Pacific itself. The enginehouse was torn down in 1952.

Steam operation came to an end on the Stirling City Branch on May 7, 1953 with No. 2528 doing the honors. Diesel units had been in regular use for some months but the upbound train of 45 empty flat cars and 5 gondola loads of sand for Pacific Gas & Electric was too much tonnage for the diesel. No. 2528's helper duty on the rear end of the train was steam's last stand on the branch, almost a half century after the first train arrived in Stirling City.

Following the closure of the Stirling City mill on January 31, 1958, the last lumber train came down the hill behind Baldwin diesel No. 5271 on February 5th. Train operation on the branch then became irregular. The only freights to run into Stirling City consisted almost solely of battered Southern Pacific gondolas which the logging contractor loaded with logs from his cold deck. The logs were then shipped to Diamond's sawmill at Red Bluff. This train operated daily, five days per week, but was limited to about two months during the late fall and early winter when the logging contractor had finished logging for the season. During the other months train operation was an occasional carload on inbound freight to Paradise. More frequent runs were made by the Chico local to switch two plants – Jessup Door Company and Butte Creek Rock Company – both situated on the lower end of the branch at the edge of the valley floor near Chico. Up until the early 1960s Butte Creek Rock Company provided heavy tonnage in shipments of railroad ballast.

Stirling City again saw freight trains on regular basis for a decade following the opening of the stud mill in February 1964. But by 1974 the future of the branch was bleak. The stud mill has been closed and Diamond announced that no further logging will be done in the Southern tract for some years. The branch's only remaining business by 1975 was the two plants near Chico.

Southern Pacific's Stirling City Branch in pre-diesel days. Extra 2312 pulls into Paradise under a cloud of smoke and pauses to switch before continuing the run to Stirling City on May 30, 1947. (Top and bottom photos by Robert Hanft; middle photo by Ted Wurm)

On June 6, 1951, the Pacific Coast Chapter of the Railway and Locomotive Historical and the California-Nevada Railroad Societies jointly sponsored a railfan excursion trip over the branch to Stirling City. No. 2728 powered the five-car special, shown at S. P. depot in Chico (above right), the Doon water tank (above), Stirling City (right), and along the line (below). (All photos by Fred Stindt except top right by Francis Guido)

The former Butte County Railroad's Barber enginehouse as it appeared in the late 1940's. An S. P. flanger and water car, used on the Stirling City Branch, are parked outside. (Bert Ward)

The final steam operation on the branch was this helper turn on May 7, 1953. No. 2528 was pressed into service to help the diesel fifty cars ahead. (Robert Lowry)

A downgrade lumber train derailed above Magalia on December 1, 1949. Section crew under Elmer Kennedy repaired the track the following day. (Elmer Kennedy)

Diesel power on the line: (Above) the first diesel "up the ridge" during its return trip through Magalia on March 1, 1950. The trial run had an unusual car on the end of the train (above right) – a business car occupied by Mr. Jennings, Sacramento Division Superintendent. (Both photos by Elmer Kennedy) The second diesel run was a year later in April, 1951. New Baldwin model AS-616 No. 5236 proved this type of diesel was well suited for the steep grades. (Elmer Kennedy) The usual diesel power on the branch for almost a decade was heavy-duty Baldwin model AS-616 units. Nos. 5262 and 5276 (below) shown on January 26, 1963. (Both Kent Stephens)

CHAPTER TEN

Meter Gauge Logging

SURVEYS FOR A logging railroad out of Stirling City started in 1902, with construction work beginning in 1903. The standard gauge railroad left Stirling City and dropped into the canyon of the West Branch of the Feather River, then climbed the edge of Bald Mountain. Grades were 4 per cent with the loads and switchbacks built to hold down construction costs. The grades also came in handy for controlling runaways, as did the adverse one into Stirling City from the West Branch trestle.

Logging started in 1904 over the first part of this railroad, with 4-4-0 No. 2 of the Butte County Railroad providing the motive power. The little American Standard received some help in its log train chores during the 1905 season when Diamond's first Shay arrived. This was a two-trucked wood burner, which became No. 101 on the roster. The second Shay was added to the roster in 1906 when No. 102 arrived. The local press proudly noted the three-truck diamond-stacked locomotive was modern in every respect.

The meter gauge equipment inherited with the purchase of Sierra Lumber Company played an important part in the decision that the logging railroad would change gauge before entering the Little Kimshew country. The basis for this decision was the high cost of attempting to put a standard gauge line into this rugged area. Unlike most of Diamond's Southern Tract — which had a volcanic ash overlay — the Kimshew was granite country: rough, vertical and full of rattlesnakes. Running a meter gauge line into Kimshew had the advantage of cheaper construction costs, since the narrow gauge could negotiate sharper curves around the rock formations than could a standard gauge line. It was expected that the company would stay in the Kimshew country for approximately a decade. Thus it was cheaper to transfer the logs to the standard gauge at some point along the line than to build it entirely standard gauge.

A site for the transfer was selected 9 miles out at Nigger Point on the side of Baldy. The standard gauge could have been built two miles beyond to Breakneck Canyon before construction costs became excessive, but the site selected for the transfer had the advantage of requiring little grading. A log transfer was installed utilizing a rollaway and two small cranes. The usual length for logs hauled on the narrow gauge was 16 feet, but 18-foot logs could be handled if they didn't overhang the brake platforms on the short cars. Two loads from the meter gauge equaled one carload on the larger standard gauge.

The meter gauge railroad by late 1909 reached Ramsey Bar, 17 rail miles from Stirling City. Ramsey Bar became a busy place: it was the headquarters camp for the Kimshew show and also the camp for the railroad crews. The Stirling Merchantile Company opened a branch store there, and undoubtedly it did a thriving business considering the Kimshew show usually operated six camps — each with 100 men.

Two of the former Sierra 0-6-0T saddletankers provided the initial motive power on the Kimshew railroad. A third saddletanker was transferred in 1910 from Lyonsville, cutting the latter logging line to one engine. Diamond's Board of Directors in September 1910 authorized purchase of two new 0-6-0T saddletank engines for Kimshaw. The new engines' most unusual feature was that Mr. Fairburn designed them to be usable on the the then-operating meter gauge and adaptable for use on the standard gauge.

The new saddletank engines arrived February 1, 1911, on flat cars at Barber. The master mechanic, James Chambers, looked at the new engines and refused to unload them — saying they were too top heavy

Building the first logging railroad in 1904. The construction crew has placed the foundation beams over the West Branch of the Feather River. (Milbourne Hedge, Sr., collection)

A "Whistling Barber" donkey roading logs from a chute to a railroad landing on the side of Baldy in pre-Kimshew days. (D. I. Corp. collection)

A donkey brings a trail of "dogged" logs down a chute. The logs are chained together with a dog on each end. This is probably the Kimshew area. (D. W. Gutscher collection)

for safe use. He was overruled. They were unloaded, inspected by the Engineering Department, and shipped to Stirling City. They started service during the 1911 season, and proved Mr. Chambers completely wrong. The most powerful engines on the meter gauge, they hauled trains of sixteen to twenty loads to Transfer Camp on the 2 per cent adverse grade, against the maximum of six to ten loads for older Sierra engines. In 1912 when the Lyonsville mill was closed, the one remaining former Sierra engine went to the Kimshew railroad. The former Sierra engines were numbered 1 to 4 on the roster, while the two new Porters became Nos. 5 and 6.

The total length of the meter gauge at the height of its operations reached approximately fourteen miles. A 1913 estimate of total mileage, including spurs, was forty-four miles. The rail weight was 45 pounds on the main line and 25 pounds on the spurs.

The Kimshew railroad ranked among the most hair-raising rail operations ever built. The cars from the Sierra Lumber Company were nothing more than 16-foot long log bunks; these short cars were connected by a center pole between trucks. Supplementing this equipment were seventy similar cars built by Diamond's Engineering Department for the 1910 season. All cars used link and pin couplers and had a platform with a hand brake. None had air brake equipment, nor did the little saddletankers – which boasted only a lever brake and a steam jam to help control the trains. Today's generation cannot help but admire and wonder at the bravery of the crews who railroaded over the mountain line, operating trains without the benefit of air brakes despite grades as steep as 4 per cent. Moreover, the main line's contours were minor when compared to the spurs, some of which climbed grades of 6 per cent or more.

The little saddletankers could push no more than three or four empty cars up the steeper spurs without having them buckle. Crews dropped the loaded cars down the spurs by gravity, since operating rules banned coupling in the locomotive while on the downgrade. The brakemen rode the loaded cars. There was one man to a car unless two cars had brake platforms placed together; then one brakeman could control two cars. The conductor and brakemen would set up as many cars as they could handle and start down the hill – hoping they would arrive in one piece at the bottom. If a crew member thought they were going too fast, he would throw his pick handle at the nearest man ahead to let him know he was "joining the birds." Desertion by one man could be disastrous while on the hill. Many remained with a cut of cars even though they negotiated curves a bit faster than they wanted.

Logging crew and donkey at Camp 1, near Ramsey Bar in 1915. (Richard Colgan collection)

A "Whistling Barber" donkey and part of its crew during the Kimshew show. The top cable was the haulback line, the bottom cable was the mainline. (George Kahn photo, D. I. Corp. collection)

First No. 6 while it was in service on the logging railroad. (Florence Styles collection)

Cuts of cars occasionally would get away, resulting in wild runaways. In 1910, Fred Anderson and two other crew members were dropping three cars down the mile-long spur from Camp One to Big Kimshew Creek. They lost control of the cut and "joined the birds" just above Camp Three. Hiking down the track behind the runaway, they found it stopped with two cars remaining on the track and the lead car over the hill. Another crew on November 21 that year wasn't so lucky. The men were letting ten cars down a spur on a 4 per cent grade with the locomotive following but not coupled. The axle on the second car broke, ditching seven cars and killing the conductor and a brakeman.

"Cornfield meets" between trains occasionally enlivened life on the narrow gauge. One resulted when engineer Charlie Marion couldn't hold No. 1 coming down the 4 per cent hill to Transfer Camp with a heavy train of logs. Unknown to the crews, a switch had been left lined for a siding occupied by a standing train. The runaway rolled into the open switch and smashed into the standing train. Crew members escaped with minor injuries, but damage to equipment was extensive. Another cornfield meet occurred when a mix-up of orders caused Nos. 5 and 6 to come together near a trestle. One train was loaded outbound and the other was returning with empties. The crews saw each other, but not quite in time to stop. Damage to the engines was slight but skeleton log cars were scattered all over the hillside.

Another potentially more dangerous incident occurred during 1910 when a Sunday excursion party came very close to being in a "cornfield meet." The excursion party consisted of fifteen people from the

Chico area hosted by Logging Superintendent Frank A. Thatcher. All went well during the ride over the standard gauge to Transfer Camp and then over the meter gauge to their destination of Camp Two. After enjoying a lunch they started homeward aboard a flat car coupled behind three carloads of logs, with No. 3 on the headend. Near Transfer Camp the engineer lost control of the train at the beginning of the 4 per cent grade that continued all the way into the camp. The crew jumped when it became apparent that they couldn't control the train. Superintendent Thatcher saved the lives of the excursionists by uncoupling the flat car and braking it to a stop. The locomotive and three cars of logs continued their wild rush down the 4 per cent grade to Transfer Camp. There they collided head-on with another saddletanker and its train. The fifteen excursionists included five young women teaching in the Chico public schools. One described her experience aboard the runaway train to a reporter:

> I never had such a wild ride in my life. We fairly flew along the road. Trees dashed by with lightning like rapidity.
>
> I did not realize anything was wrong and thought we were just making time until Mr. Thatcher with a ring of terror in his voice shouted to us to hold tight and not jump.
>
> Then I noticed how terribly fast we were going. Men were tumbling off the front of the train and rolling in the dust as we swept by. Mr. Thatcher was the hero and if he hadn't crept up in front of the swaying car and uncoupled it I'm sure some of us would have been killed. He hung on the brake and twisted it with grim determination. Slowly, slowly the car commenced to slow down. Just ahead of us the train was tearing madly down the grade. Then there was a crash, the engines had collided with a deafening roar. It was exciting while it lasted, but I don't want another such experience. The men told us after it was all over that we women behaved admirably. I guess it was because we were too scared to do anything else.[1]

Both saddletankers were repaired and returned to service.

Logs frequently fell off cars, causing a delay whenever any that failed to clear the train had to be pulled free. Each engine carried a roll of heavy cable behind its fuel tank. Called the "jerk line," the cable pulled logs clear of the train and helped re-rail cars. Logs that fell clear of the train, as well as those pulled clear by the crews, were left for a special clean-up train which ran regularly each fall at the end of the logging season. A Dolbeer donkey on a specially designed heavy flat car loaded the logs lost along the railroad and salvaged cars which had jumped the track. The immense size of this annual fall cleanup can be visualized by the fact that the job often took as long as a month.

Snow problems also distinguished the Kimshaw railroad, and there was an effort every spring to clear the tracks for use. While every year brought a struggle, the company's decision to "begin early" in April, 1913, brought even more difficulties.

There had been heavy snows during the winter which continued until April 1. The Sacramento Valley enjoyed warm weather for several weeks, resulting in several hundred idle men waiting in Stirling City for the season's work. On April 3, a crew went to work excavating the roundhouse, which was still buried under a ten-foot blanket of snow. The job of "breaking through" started the next day, involving a crew of several hundred men. One Shay, equipped with a steel

A meter gauge train at the transfer. The two nearest cars have their brake platforms together, a measure to help control cuts of cars on spurs. Standard gauge cars were on the other side of the rollaway. (Mrs. Eda Seaman Scott collection)

Left: The meter gauge line winds through the rough Kimshew country on its way to Ramsey Bar. (Richard Colgan collection) Below: A "chute greaser" and his "boat." The "boat" was hooked to the last log on the trail; the greaser rode the boat, swabbing tallow on the chute enroute. (Mrs. Eda Seaman Scott collection)

plow, started on the standard gauge to Transfer Camp. Then the crew started clearing the meter gauge. The snow removal equipment consisted of a steel plow mounted on the front of the "one-spot" – and plenty of shovels. The average depth of snow was five feet, and it some places it drifted up to nine feet on the level. The engineer would back No. 1 approximately 300 years, put it in forward motion, and open the throttle wide, hitting the snow bank at full speed as a plume of snow cascaded over the engine. Most of the time the little saddletanker would ram her way about 25 feet before stalling – unable to go forward or back. Each time the engine stalled, the shovel gang dug her drivers free so No. 2 could pull her out of the snow bank. After several weeks of this back breaking toil at an average cost of $225 per mile, the railroad finally opened and the crews turned their energy to excavating the camps.

Troubles were not always confined to the meter gauge during those highball logging days. Three standard gauge locomotives figured in a strange series of accidents that happened on one March day in 1910:

> Good Friday was certainly an unlucky day at Stirling City. Commencing early in the morning and lasting through the entire day there was naught but one string of accidents, but these resulted fortunately in no one being seriously injured . . .
>
> To begin with J. W. Mutart, a lumberman, got his leg caught in a sniping chain and before the engineer on the 'donkey' could check the cable it had wound around and broken the limb of the unfortunate man. He was removed to Parker station and from there to Chico. The accident happened at Camp Three.
>
> The next thing that happened was for locomotive 101 to break its axle while on one of the logging roads. This, of course, tied up the train and engine No. 6* was sent for from Stirling City. Number 6 arrived O.K., gathered up the broken parts, and started back to Stirling City. On the way the rails spread and the engine was thrown into the ditch. The men jumped and no one was seriously injured although it is reported that several received slight bruises.
>
> These two engines being out of commission, it was necessary to send to Chico for another locomotive and engine No. 102 was dispatched from Barber. On the way to Stirling City the locomotive ran into the platform at the new station house below Paradise and tore off the platform and partially wrecked the building.[2]

*Parker Station that the injured logger was taken to was on the Butte County Railroad. No. 6 was Diamond's first No. 6, the standard gauge Porter 0-4-2 inherited from the Sierra. Within a year after the accident, it was renumbered No. 103 when 2nd No. 6 was purchased for the meter gauge.

With one logger, three engines and a depot out of commission they were probably glad to call it a day and go home, hoping the next day would be better.

Unusual problems were not confined to the logging railroad. Someone decided during the 1912 season that it would be faster to use a skyline at the Stirling City pond to pull the entire load of logs off a car at once. It was also decided not to take time to properly deck the logs, so a huge jumbled pile of logs built up, quite unlike the usual orderly stacked logs in a cold deck. The decision not to properly deck the logs proved to be a costly mistake. Crews had a terrible time pulling it apart, especially after it froze solid in late fall. They used everything they could find to pull the logs out of the "deck," even locomotives. Many logs were sinkers that promptly went to the bottom of the pond when pulled free. They tried this method of decking logs only once!

The highball logging days ended with the Kimshew Show in 1915. The last trainload of logs reached Stirling City on November 1 and work started immediately on tearing up the tracks. When the dismantling train reached Transfer Camp, Diamond was out of meter gauge logging forever. Beginning with Frank Thatcher's contract logging in 1916, all logging railroads were standard gauge. The logging production was greatly curtailed at the end of the Kimshew Show, and its production figures were not to be matched until the early 1950s. But the huge crews of the Kimshew Show were gone forever from Stirling City.

Homemade crane, with a small donkey engine and wood boom, at the transfer. Logs were dumped from meter gauge cars by the crane. Will Thatcher, transfer foreman, at right. (George Kahn photo, D. I. Corp. collection)

A portion of Transfer Camp. Cookhouse on the right behind the sheds. Water was obtained from an excellent spring and pumped up to camp by a steam pump. (George Kahn photo, D. I. Corp. collection)

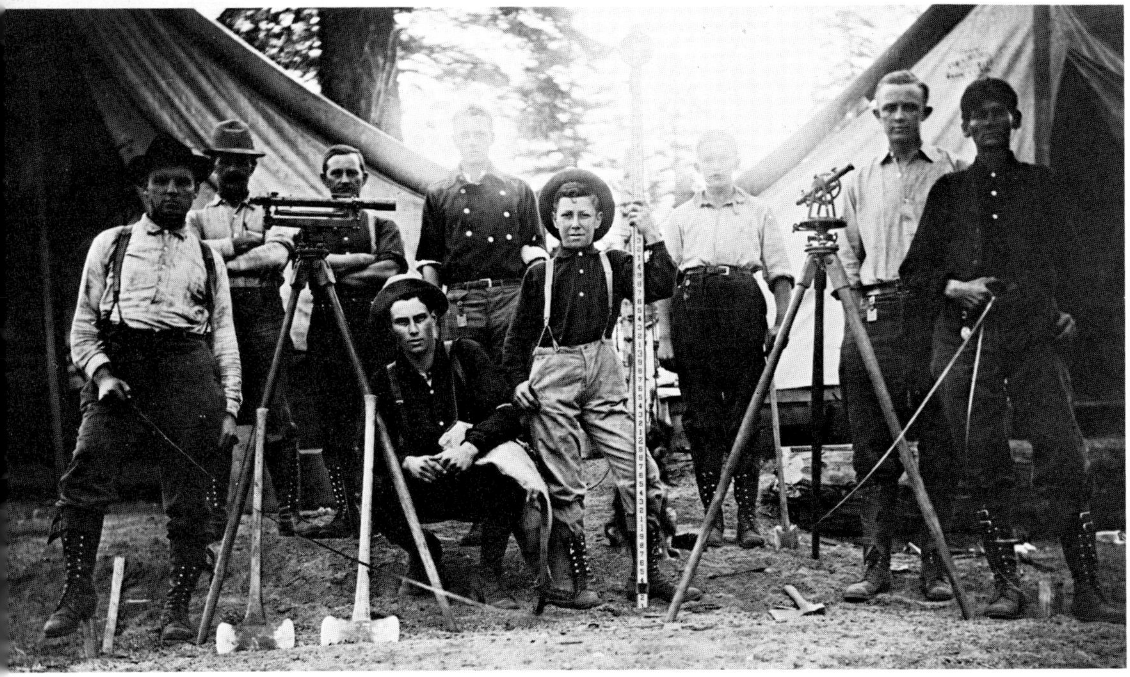

A survey crew at one of the Kimshew camps. The tallest man on the crew (fifth from left) was Frank ''Comp'' Compton, who later became superintendent at Stirling City. The man at right was Bill Stokes, who joined Diamond at the very beginning at Stirling City. (George Kahn photo, D. I. Corp. collection)

Meter gauge car shops at Transfer Camp. The carpenter shop is on the left; the building in background was the wheel and brake rigging shop, which also serviced locomotives. Next to it a crew shovels off ballast. Note the stub switches. (George Kahn photo, D. I. Corp. collection)

Shay No. 101 pauses to take water at a tank next to the West Branch of the Feather River. (Mrs. Justa Farley collection)

Kimshew logs on standard gauge cars at Stirling City. Two meter gauge loads made one standard gauge load. Note that the track is inclined toward the log rollaway and pond. (Mrs. Eda Seaman Scott collection)

The Engineering Department designed this compressed air log unloader to dump Kimshew logs on the rollaway next to the pond. (U. S. Forest Service photograph, D. I. Corp. collection)

No. 102 and crew pause for a photographer during Diamond's early years. The engineer is believed to be E. L. "Al" Myers. Crew members are unknown. (Mrs. Justa Farley collection)

A Kimshew log hits the pond; until 1915 the dump was on the town side of the pond. Peaked-roof building was the standard gauge enginehouse. (Mrs. Eda Seaman Scott collection)

A fourteen-car train behind No. 6 crosses the "Ramsey Bar" trestle spanning Kimshew Creek. (Mrs. Eda Seaman Scott collection)

This meter gauge logging trestle illustrates why logging was so difficult in the rugged Kimshew country. (D. I. Corp. collection)

Saddletanker No. 4 and the crew wait at a junction on the Kimshew line. (Fred Anderson collection, D. I. Corp. files)

Moving camp at Ramsey Bar in 1912. The cabin nearest the saddletanker has just been loaded. (Richard Colgan collection) The cookhouse at left was presided over by Mrs. Eudoxie Green, head waitress in Diamond's camps for years. The hard-working Kimshew loggers ate their meals in these tent cookhouses, which must have been cold in the spring and fall. (George Kahn photo, D. I. Corp. collection)

It's early morning at Ramsey Bar and crews and engines are lined up before beginning the day's work. Note the contrast between No. 6 and the older saddletankers. (Frank Compton collection, D. I. Corp. files)

Apparently it was decided that No. 6 should be closest to the camera for a second photograph, so the entire sequence was reversed (below). The cable on each tank is the "jerkline." (Douglas Richter collection)

The Kimshew line had its share of wrecks. Saddletanker No. 3 has turned over (left), and a crew is on the scene to set it back on its wheels again (below). (Both Florence Styles collection)

You can almost hear this meter gauge car groaning under the weight of a huge 16-foot diameter red fir butt log. Note the brake handle and platform. There were no air brakes on the meter gauge. (D. I. Corp. collection)

The winter log deck at the end of the pond in 1912. The experimental "deck" proved to be an ineffective method – it was virtually impossible to pull logs out of the frozen pile. (D. I. Corp. collection)

The log pond held five million board feet of logs in this 1910 view from the mill. The buildings across the pond include the warehouse, railroad car shop, and enginehouse. (D. I. Corp. collection) The picture below shows Ramsey Bar just before the meter gauge ceased operations in 1915. (Richard Colgan collection)

CHAPTER ELEVEN

Rails through the Pines

THE WHEELER TRACT north of Stirling City was purchased to provide Thatcher with timber for a few years. In August, 1915, construction started on a new standard gauge logging railroad north of Stirling City. The logging railroad was finished during the winter months, along with a logging camp just four miles from town.

The Thatcher contract era logging made a great contrast to the Kimshew show. During Kimshew, up to 300,000 board feet of logs hit the mill pond each working day. In 1916, Frank Thatcher's logging camp put in about 100,000 feet daily, with a corresponding drop in the total season's production from the 40 million board feet averaged each season during Kimshew. Thatcher's daily log production was hauled over the 5½ mile standard gauge railroad by saddletankers Nos. 5 and 6, both converted to standard gauge at Thatcher's expense. Neither Shay saw service during the 1916 or 1917 seasons. By the 1918 season, however, the logging railroad had been extended to seven miles; all four engines rolled in service for the first time since 1915. The engines and all 72 cars were equipped with train air.

By 1917, Thatcher, along with many other West Coast loggers, began to experiment with a form of "high lead" logging in an attempt to get his logs off the ground. Until then, he logged by "ground lead," as did many other companies. They used chutes and donkeys to move the logs to landings. High lead logging proved a great improvement over ground lead, in which the logs were dragged over the ground – where they frequently hung up on projecting stumps and rocks. High lead logging utilized a spar tree and aerial cable lines to get the logs off the ground.

Thatcher's first experiment was to place a ladder against a spar tree, on which a haul back line was hung approximately 16 feet off the ground. In 1918 he went into a modified high lead that pulled logs as long as could be handled by the donkeys. This modified high lead system, employing only a mainline and a haulback line, was used for two years until 1920. At that time, Thatcher went to a true high lead.

Diamond purchased new logging equipment in 1920, adding two Willamette two-speed logging donkeys. Another Willamette donkey with a larger three-speed engine was purchased the following year along with a McGiffert log loader and skidder. The McGiffert, named for the Minnesota logger who invented it, was popularly known in logging terminology as a "jammer." The McGiffert's machinery was situated on an elevated deck, which was mounted on a pair of railroad car trucks. The jammer, self-propelled, moved itself along the railroad to new loading sites. Once at a loading site, the jammer was placed in its loading position by the lowering of the four corner posts of the machinery deck. The corner posts, known as "spuds," were each equipped with a broad shoe on the bottom. It rested on the ties outside the rails. As the machinery deck was raised the trucks also were raised to a horizontal position under the deck. Once the jammer reached loading position, it pulled the empty logging flats underneath the machinery deck and loaded them in front via a 40-foot boom. The jammer thus could load anywhere on the mainline or spurs without need of additional trackage.

Thatcher finished logging the Wheeler tract in 1920. This was ahead of schedule due to increased production demands. The logging railroad extended toward Inskip as logging moved onto Diamond's own lands. Construction costs on the mainline and spurs ran as high as $25,000 per mile; $5,000 of this was surveying cost.

Dana Bailey, Diamond's "high climber," sits atop a spar tree after topping it. The crowd below consists of retail yard managers and guests gathered for their annual meeting in 1919. (D. I. Corp collection)

The Thatcher contract ended with the 1922 season and Diamond went back to operating the mountain show itself. Replacing newly appointed Superintendent Compton in his former job as logging engineer was his older brother, C. S. Compton. He brought both his son and another young surveyor, Stanley Keesy, to work under him. Both went to Stirling City with the intention of staying a few years; Stan Keesy remained there for 43 years. During much of this time, he was construction engineer and trainmaster over the logging railroad.

During the 1923 season "Comp" had two logging camps in operation. Dana Bailey ran the camp at Inskip and Herschel Cullen directed "Bridge Camp" near Wild Yankee Mine. This season Diamond had to contend with its only labor troubles of the post World War I era of the I. W. W. lumber camp agitation. A strike called at Herschel Cullen's camp brought Dick Colgan and Frank Compton to investigate. "Comp" went well equipped for the occasion, wearing a gun and his deputy sheriff's badge. He ordered a special train made up, announced anyone wanting to quit could ride it to Stirling City, and ordered the remainder of the crew back to work. About 5 per cent of his crew rode the train to Stirling – and most of them were back on the job within a week. The two agitators were followed to town and were warned to keep going.

At the end of the season, both logging camps were abandoned and the combined equipment moved to a new camp at the Walker claim between Inskip and Bottle creeks. Cullen resigned at the end of the season and Dana Bailey ran the single camp. A few months later he was promoted to logging superintendent.

During 1924, the logging railroad was extended four miles above Inskip. This extension marked a company first with most of grading accomplished with a steam shovel. Logging railroad grading up to this time had been by pick and shovel plus Fresno scrapers.

The tragic Bull Creek trestle collapse of 1926. Crane No. 1 lies in the debris as flat car No. 608 hangs on the edge. The flat car, loaded with a small oil tank for the crane, had been shoved across just before the trestle collapsed. (Mrs. Eda Seaman Scott collection)

In 1926 the logging railroad was again extended by building from Inskip to Bull Creek. One bad accident marred this construction work. The Bull Creek trestle had just been completed. The construction train conductor signalled for the locomotive crane and 0-6-0T No. 5, behind it, to cross the trestle. The engineer on the five-spot failed to see the signal and the crane started across alone. As it neared the center, the trestle suddenly collapsed and the American Hoist crane and its two-man crew dropped 70 feet to the bed of the creek. One man was killed and more undoubtedly would have died had the engine followed the crane.

An incline spur was built from the railroad into the creek bed, and the crane was brought up from its resting place for rebuilding.

There were other accidents in the 1920s, but they were minor affairs when compared with the bridge collapse. One morning a crew forgot to take water before leaving Stirling City with a string of empty log cars. The Shay soon ran short of water, and low water in the boiler caused the crown sheet to drop. The crew was fortunate! No. 101's boiler wasn't tight and there was no explosion. (A tight boiler would have resulted in the head-end crew being sent to eternity.) The Shay was repaired and returned to service, only to be hit head-on two years later by a runaway log car that broke the engine's frame.

Most runs with any engine were made without incident, but troubles plagued an occasional trip. One in the 1920s is particularly memorable, for it holds something of a record in derailments during a single trip. No. 102 was on the ground 13 times while bringing in a trainload of logs from Hi-Line camp, above Inskip. Instead of the normal 1½ hours, the run took 23 hours. One of the Shay's trucks was cross-wise to the rails on a trestle, harassing the crew with the worst derailment of the trip. Oddly enough, not one of the log flats so much as got one wheel off the rails during the entire run.

Diamond operated two inclines during the 1925 and 1926 seasons. The first was on Rimple Ridge at the end of the extension of the mainline between Bull and Butte creeks. It ran 3,000 feet down Rimple Ridge to a donkey landing.

The Rimple Ridge incline, abandoned at the end of the 1925 season, was replaced with an incline at Coon Creek during the following season. The Coon Creek incline extended from the mainline a little less than 4,000 feet up the hill. The "big road engine" (Willamette wide-face logging donkey) served as a hoist atop the incline to raise and lower the log flats. There was about a mile of track at the top of the incline and one of the two Porter saddletankers was taken up the incline to work on top. This incline also lasted for one season.

The company in 1925 started development work on the isolated stand of approximately 250 million feet of timber on the west side of Butte Creek. Reaching this timber with a logging railroad presented a formidable construction problem. Stirling City and the logging railroad were on the east side of Butte Creek. The timber was on the west side of the creek, which wound its way toward the valley in a canyon ranging in depth from 1,200 to 1,500 feet. The timber tract was only six miles airline distance from the Stirling City mill.

Snow! Opening the logging railroad in the spring was a major project. (Top) two Shays battle deep drifts in April, 1927. (Center) No. 102's crew checks the Shay's fuel supply. (Bottom) No. 102 crosses a trestle. (All Milbourne Hedge, Sr., collection)

By mountain road it was thirty-five miles!

Several possible solutions were discussed. One proposal was to cross the canyon by building a railroad down into it, and then up the opposite side. This was ruled out due to the huge investment of building thirty-five miles of steep grade through difficult rock formations. Another possible solution was to circle around the head of the canyon through Butte Meadows and then down the west side to the timber. This would have been a rail haul of at least forty-five miles, but mileage did not tell the whole story. The railroad out of the timber would have several stretches of 2 per cent adverse grade against the loads, requiring every train to be double-headed. Several more Shays larger than No. 102 would be required for this haul. Both proposals would result in a continual heavy operational expense.

A third solution was proposed by Dick Colgan, who suggested a direct crossing of the canyon by means of a huge double incline with a span of 7,000 feet. The double incline actually was one on each side of the canyon — climbing the canyon side in direct lines tangential to each other. Loads would be lowered by the incline on the west side (actually a decline) and raised by the east side incline; the reverse procedure would be followed for the empty cars. Despite the estimated $90,000 cost of the double incline, it was far cheaper than the alternate routes in both construction and operating cost.

Plans had been prepared by 1926 in consultation with Leyland S. Rosener, San Francisco engineer, and construction work started on the eastern approach at Powellton. Building on the incline itself began in 1927, under direction of Construction Engineer J. C. Due and William C. Breedlove, Superintendent of Railroad Construction. The work on the incline was extremely difficult, due to the steep grades on the canyon sides.

No. 102 and her regular crew at Stirling City in 1927. From left: Floyd Clark, fireman; Brad Smith, conductor; Norman "Dutch" Hubbard, swing brakeman; Ed Myers, engineer; and Gene Seybold, head end brakeman. (Norman Hubbard collection)

Brand new Willamette No. 104 posed just after her delivery in 1927. From left: Gene Seybold, John Peadon, both brakemen; Brad Smith, conductor; Norman Hubbard, brakeman; Dick Ferris, conductor; and Al Myers, engineer (Ed Myers' brother). Fireman Harold Clark stands on top of the tank, and Lorene Myers, Al's daughter, stands in the gangway. The big Willamette had a series of misadventures during her log-hauling career. The first was in 1927 (below), only a few months after delivery, when she derailed while backing up to the approach to Little Secret Creek trestle and dropped partially through the bridge. The close-up (right) shows Brad Smith standing in the gangway, followed by Al Myers, Bill Breedlove, and the section boss. (All Norman Hubbard collection)

*One of the incline hoists at Vulcan's Wilkes Barre, Pennsylvania, works. (*Diamond Good Fellowship Bulletin*) At right, the double incline crossing Butte Creek Canyon looking from the east side. (Dana Bailey collection)*

Following the grading of the roadbed, track in 30-foot sections was lowered, one section at a time. A 65-foot bridge was necessary to cross Butte Creek at the bottom of the incline. The steel girders for it were placed by a skyline.

By mid-1927 the machinery was arriving for the incline. The Vulcan Iron Works built the electrical hoist machinery, and ten rail cars were required to transport the 200,000-pound hoists and the other machinery from Wilkes-Barre, Pennsylvania. The hoist on the incline side at Powellton (i.e., the hoist pulling the loads up) was the larger of the two hoists. It had a ten-foot drum with 4,400 feet of 1¾-inch cable. The smaller west side hoist had an eight-foot drum carrying 3,400 feet of 1½-inch cable. A 900-horsepower Westinghouse motor powered the big hoist, and a 400-horsepower motor operated the other hoist. Power came from a 60,000 volt line built by Pacific Gas & Electric Company. The motors were regenerative: the descending load on one side of the canyon generated electric current to supplement that supplied by the power company. The Vulcan hoists could raise and lower 75 tons, which was more than adequate for a direct-line incline handling one car at a time.

The building of the incline made it necessary to devise some means of holding the logs on the cars while they were on the steep grades. Instead of bulkhead-equipped flat cars used on the Yosemite Lumber Company's huge incline, Diamond chose another method. Instead, two different attachments devised for the ends of each car enabled both loaded and empty cars to be handled on the incline. Strips of one-by-six iron approximately four feet long were hammered to form a horn at each end. Two of these were installed on the ends of each car. A cable line was placed around the logs near the center of the car and then the ends of the line (called chokers) were run through the horns to form a "V," extending beyond the end of the car. The main cable line, hooked directly to the ends of the chokers, caused the cable on the cars to form a kind of a "bridle" to hold the logs tightly. This was in addition to the usual cables or chains holding the logs on the cars.

The empties were handled on the incline by means of an iron plate, bolted to the main timber at each end of the car. The plate, parallel to the main timber, crossed the horns at a right angle. The end of the plate above the coupler was hammered up to form a U-clevice. A pin or bolt hooked the main cable line into the clevices. This method was used for only one season before it was discovered some of the U-clevices showed signs of metal fatigue and were starting to break. They were replaced with two pieces of angle iron bolted to the main timber of the car in the same location. A hole was drilled in each piece directly opposite to each other. A heavy bolt was inserted through the holes and the main cable line hooked directly to it.

The east side incline had several extreme variations

The Butte Creek double incline three years after construction: west side at left, east side at right. By this time, the extreme variations in the east side grade had been reduced. (Diamond Good Fellowship Bulletin)

The east side hoist operator had this view of the large drum. At right is car 669, showing the "horns" and U-clevice designed for crossing the incline. The U-clevices were replaced after one season with the strips of angle iron. (both Richard Colgan collection)

in grade which did not occur on the west side. Immediately after leaving the edge of the canyon, the track plunged into an increasingly steep grade, reaching a maximum of 62 per cent. This continued for 1,500 feet and then the track rose out of a little sag to cross the P.G. & E. power ditch on a trestle. The grade then eased to 10 per cent for 1,220 feet before nosing over into a 65 per cent grade that continued for the last 1,000 feet of the incline to the bottom. Company employees were occasionally permitted to ride the incline: at this point they would be lying flat on their back, their feet braced against a cross timber while they looked straight into "the hole." Grades on the other side of the incline reached a maximum of 58 per cent without the extreme variation of grade on the east side.

The "flat" below the power company ditch on the east side had several operating problems. When empty cars were lowered, the hoist operated at a steady speed of 600 feet per minute, but the car slowed to approximately 300 feet per minute on the flat. Slack cable built up behind the car and snaked all over the ties as the car crawled down the grade. As the car nosed over into the 65 per cent grade, it would pick up speed in excess of the cable speed while it took out the slack in the cable.

Another problem with the "flat" occurred as soon as the car was on the 65 per cent grade below it. The cable became taut between the hoist house on the edge of the canyon and the track at the beginning of the heavy grade, rising several hundred feet in the air above the track on the flat and the sag above it. Company officials decided to control this extreme rise by means of a hold-down tower built in the middle of the "flat."' The hold-down tower consisted of an I-beam with rollers mounted on two posts high enough to clear the loads.

One new locomotive was needed for the new logging operations, and this was a 70-ton geared locomotive for the run on the east side between Stirling City and the incline. A 70-ton Shay type geared locomotive was purchased new from the Willamette Iron & Steel Works at Portland, Oregon. One of Diamond's locomotive engineers, Al Myers, served as engine messenger to bring the engine to Stirling City and then became its regular engineer. The new Willamette became No. 104 on the roster.

The incline began operation during the 1928 season, and the previous logging operations near Bull Creek were abandoned. Initially the new camp on the west side of the incline was Logging Camp No. 1, approximately four miles up the track from the west side hoist house. It operated for one season and was abandoned for a new camp, Logging Camp No. 2 at West Branch, which remained the main logging camp during the rest of the time the incline was operated. Two locomotives were taken across the incline to the west side. Shays Nos. 101 and 102 became the west side engines, with No. 102 doing most of the road work and No. 101 switching the spurs. The 101 also served as helper engine on the mainline, shoving on the rear of the log trains on the two per cent adverse grade out of West Branch to the summit of Carpenter Ridge.

Mr. Colgan succeeded "Comp" as superintendent in October 1928, just in time to contend with a bad forest fire and the first wreck on the incline. The wreck, a rather spectacular one, occurred when the east side hoist operator failed to pull a car far enough over the "holding hump" at the top of the incline. The cable was taken off and the car started back over the hump. Picking up speed, it took off down the incline. Within seconds it sped so fast down the 62 per cent grade that it was airborne. The car sailed completely over the P. G. & E. ditch trestle and landed with a thunderous crash to the left of the incline below the ditch. It landed, virtually demolished, upside down. Its wheels were several hundred feet away. The logs, somewhere in the mid-air flight, broke loose from the car. Most of the 60,000-pound load of logs continued flying through the air for another 300 yards — scoring a bulls-eye hit on the hold-down tower. This structure suffered the same fate as the car, being demolished so completely that the only thing salvaged was one roller.

The damage repaired, operations resumed without a hold-down tower. A few weeks later came the second runaway wreck, directly traceable to the first. Mr. Colgan was in the east side hoist house, when both he and the hoist operator saw the cable suddenly go slack while a loaded car was being pulled up the 62 per cent grade directly below the top of the incline. Mr. Colgan quickly pushed the warning button to alert the crew in "the hole" to the danger. The hoist operator shut off the power. Seconds later they heard the thundering crash landing of another airborne car and its load.

The second wreck was caused by one of the choker cables used to hold the logs on the car. When the car reached the flat below the power company ditch, the hoist cable started settling *down* faster from its hundred foot height than the 600 feet per minute that the car was being pulled *up* the incline. The choker on one side bulged — possibly because it may have been new, stiff cable. Then on the pull up the 62 per cent grade above the ditch, the choker pulled taunt against the front car wheel on its side. Within seconds the car

Four views of Camp No. 1, first logging camp on the west side of the incline: 1. Two-man buckhouses; 2. A camp "street"; 3. Camp cookhouse, feeding 120 men daily; 4. Crew going to work in the morning. (Diamond Good Fellowship Bulletin)

It is obvious that this spur in West Branch Camp would never carry the weight of a log train. The two-man cabins on each side of the spur were brought in on flat cars, the only reason for the lightly built spur line. (Cliff Gilbert collection)

Sequence of moving logs from the West Branch camp via the incline to Stirling City: (Top) Shay No. 102 works up the 2 per cent grade of Carpenter Ridge to the incline with (Center) No. 101 shoving from the rear. (Both Cliff Gilbert collection) Below is a car being eased over the edge of the west side incline. The "bridle" holding device can be seen here. (Norman Hubbard)

wheel ground through the choker cable. That turned everything loose!

Once again the car travelled so fast it became airborne before reaching the trestle. It jumped over the trestle and landed where its predecessor had a few weeks earlier and "scattered around." The major difference was the first car missed the incline, but the second one and its entire load came down on the incline track. It took several days to replace more than a hundred feet of broken ties and twisted rails before production could resume.

Officials decided not to take any more chances with loose chokers, so "Dad" Burgess was taken off his job in "the hole" and posted with his lunch bucket and chair midway on the "flat" to visually check each car as it passed him. If anything was amiss, he signalled the hoist house with the bell line. There were no more wrecks due to loose chokers.

"Dad" Burgess spent less than a month at his post before the operations closed for the season. Before the 1929 season started, the problem that caused the wrecks was eliminated by raising the grade on the "flat" to 15 per cent from 10 per cent. This kept the cable taut and the east side incline gave no further trouble.

The incline was a dangerous operation, but there were relatively few accidents on it and no fatalities during its eight years of service. Besides the two cars wrecked on it, the only other equipment piled up were two logging donkeys.

Construction started in 1935 on the mainline to Butte Meadows, site for a new logging camp following the closing of the West Branch camp. The new mainline as far as Bull Creek used the grade originally built in the 1920s and abandoned when the incline opened.

The West Branch camp was abandoned in July, 1935. The equipment and flat cars loaded with the cabins moved across the incline on July 4 to Powellton. An interim logging camp established at Powellton remained in operation until the fall of 1936.

Shay No. 102 remained at West Branch after the camp was abandoned. By this time, the rails did not have to be hauled across the incline as when West Branch was opened. One of the numerous spurs out of West Branch had been extended several times until it reached the lower end of Butte Meadows over roughly the same route of one of the proposed alternates to the incline. The spur was left intact so the rail on the west side could be hauled directly to Butte Meadows and stockpiled for spur construction.

No. 102 and a crew started pulling up the west side mainline, working from the hoist house to the camp. After all spurs were pulled, the crew worked their way up the spur to Butte Meadows. No. 102, arriving at Butte Meadows before the mainline was finished, was stored for several months on the end of the spur. She was regularly used for rail pulling chores, as the wheel tread on the Shay was almost an inch wider than the wheel tread on the other engines. This wide tread enabled No. 102 to stay on the rails after most of the spikes had been pulled. The narrower tread wheels of the other engines would drop onto the ground as the rails spread.

The incline and hoists were removed during 1935 and 1936. The steel bridge at the bottom of the incline was raised from its foundation and skidded up the east side incline to the top. The rails were removed behind it. The incline hoist machinery and the bridge were stored pending re-use: the logging railroad was torn up back to the mainline after the Powellton camp was abandoned.

The Butte Meadows camp, opened near the end of the 1936 season, remained longer in one location than any other logging camp. The logging railroad mainline remained unchanged during the remainder of its operating life, although spurs were built out of Butte Meadows. It was also under the same operating supervision during this entire period, as Stan Keesy became trainmaster over the entire railroad in 1935. Prior to that he had been west side trainmaster.

There were heavy rains and snow during the winter of 1936-37, and the short supply of logs at the mill resulted in the decision to open the railroad early. The depth of the snow was reduced by rains in March, which enabled the railroad to start plowing out the mainline.

The plow train started out of Stirling City with No. 104 on the point pushing the plow and No. 102 be-

Changing cables in the "hole" at the bottom of the inclines. The loaded car at right has arrived from the west side, the empty flat from the east. The cables will be switched on the two cars, which will then continue their respective ways. (Norman Hubbard) The car (below) nears the top of the east side. (Richard Colgan collection)

Diamond's railroaders battle deep drifts about one mile below Inskip as they struggle to open the mainline to Butte Meadows in March, 1937. Nos. 104 and 102 (above) backed up to allow sectionmen to dig snow from around the wedge plow. No. 104 (below) in trouble a few miles beyond the cuts. It took the combination of a spar pole, a crane, and a Caterpillar tractor to raise her back to the track. The spar pole placed in position (above right) and (below right) the engine raised to track level. (All Norman Hubbard)

hind. Al Myers was at the throttle in No. 104's cab while Norman "Dutch" Hubbard fired the Willamette. Back in No. 102's cab, Ed Myers (Al Myers' brother) was on the right hand seat. The combined power of the two locomotives pushed the plow steadily, through the drifts until mid-afternoon. Then they had trouble: several cuts between old Y Camp and Swede Cut had drifts higher than the engine cabs. More than once they had to back up and ram into the snow at full speed. Several times the plow had to be dug free by the section crew, and then it derailed while they were backing up. The derailment was due to the usual cause – piled up snow behind the blade lifted the front truck off the rails. (This was later cured when the car shop devised a hydraulic air ram that enabled the blade to be raised.) The plow was finally rerailed and they resumed their battle with the drifted cuts until calling it a day and backing to Stirling City to tie up for the night.

Next morning, they again battled the drifted cuts. Shortly after lunch they broke through the last drift and moved into light snow. But just ahead lay disaster. They had covered about a mile when the plow train reached a small fill. The train stopped while Dana Bailey and several other employees walked across the fill and carefully inspected it. It looked and felt solid despite the wet roadbed – there was little snow on the fill as it had good exposure to the sun. But the fill's appearance was deceptive: it was honeycombed underneath with soft spots from constant freezing and thawing. They signalled the plow train to start across. As No. 104 neared the middle of the fill, it suddenly started to slide out on the lower side. Al Myers shut off steam and Dutch Hubbard jumped off the cab steps on the upper side and ran back to uncouple No. 102. Ed Myers hastily backed off the fill and as Al Myers joined the men assembled on the bank, they started to devise a way to save the 104.

They wisely decided not to attempt to back No. 104 off the fill. It was too dangerous and the 104 – in the words of one participant – "would have kicked the track right out from under her!" There wasn't time to recouple No. 102 to the Willamette. The couplers were already out of line, plus there was danger if the 104

Diamond pioneered the use of Dow chain saws in logging. Here is the first saw in 1933 with a group of Dow and Diamond personnel. Logging Superintendent Dana Bailey fifth from left, Dick Colgan second from right. At far right is George F. Cornwall, longtime editor of The Timberman. *(Richard Colgan collection)*

A 19-car log train moves through Powellton on the east side of the incline with Willamette No. 104 on the point and No. 102 pushing at the rear. These pictures were evidently taken at the end of the season, as No. 102 normally worked on the west side of the incline. (Both Dana Bailey)

went over suddenly, she might take the lighter Shay with her. The fill continued to slip out until the wheels on the upper side raised several inches off the rails. Then the slippage slowed and the 104 remained in that position for some minutes. The crew managed to hook a cable to the upper side of No. 104 with the intention of snubbing it to a large tree above the track. They almost had the cable anchored to the tree when a rail on the lower side suddenly broke. Almost simultaneously with the pop of the breaking rail, the 104 rolled over and came to rest upside down at the bottom of the fill.

The crew took No. 102 back to Stirling City, but a snowstorm started the following day and they couldn't get back to the wreck until mid-April. After plowing out the track, the crews started to raise No. 104. A spar pole placed above the track and the combination of No. 102, a locomotive crane, and a "cat" raised the 104 up the track. The Willamette, placed on her trucks, was towed to the Stirling City shop, returning to service when the season opened on May 4.

During the 1937 through 1939 seasons there were two major spurs out of Butte Meadows. The Lower Cascade Spur extended 12 miles into the Lower Cascade Unit from the end of the mainline at Chico Meadows. The Lower Cascade Spur crossed Highway 32 about one mile above Soda Springs by the bridge that had been used at the bottom of the incline. Three flat cars took it there and the locomotive crane lowered it into position. The Upper Cascade Spur was built in 1938 and extended in 1939 to bring the total length of the logging railroad to 38 miles.

Hollywood came to the logging railroad in 1938 when it was used on location to film "Stand Up and Fight" by Metro-Goldwyn-Mayer Studios. The story was set in the Alleghanies near the Cumberland Gap of Maryland during the early 1800s when the railroads were first being built west, displacing stagecoach lines. In the plot of "Stand Up and Fight," the stagecoach company decided to stand its ground against the intruding iron monster; the script called for a race between a stagecoach and a train as part of that battle. The site selected for filming was Chico Meadows, which M.G.M. thought approximated the scenery in the East. The train used in the film was supplied by the studio, which obtained it from the Baltimore & Ohio Railroad's historical museum. Two cars and a locomotive, a 4-2-0 operating replica of the "Lafayette" (built by Norris in 1837), were shipped to the West Coast in a mill gondola. The railroad also furnished an engineer and fireman, the latter being Lawrence W. Sagle from the B. & O's publicity department.

The movie train at Chico Meadows. Left to right: Horace Hough, MGM assistant director; Dana Bailey, logging superintendent; Richard Colgan, production manager; Rex Black, California Forest Protective Association; and Richard Rossen, MGM director. Lawrence Sagle, standing pipe in mouth on the engine deck, was in charge of the B & O equipment. (Richard Colgan collection)

The stay at Chico Meadows was not without its problems, according to Mr. Sagle's recollections.* The movie train used Diamond's logging railroad track for filming work, but had its own special spur for storage. While using the mainline, the crew arranged for the log train to stop out of camera range and signal its presence with a single blast on the whistle. Then the movie train would get on its spur as soon as the scene was completed, signalling to the log train with two whistle blasts that the line was clear. All went well until one morning when the crew forgot about a waiting Shay and its train setting out of sight around a curve. The director ordered the crew to make another run past the camera, blowing the whistle as they passed a stump. They steamed past the camera and blew the whistle twice at the marker. Immediately there were two answering whistle blasts from the waiting Shay. The Shay's engineer opened the throttle and it urged its train of logs into motion, the rapid exhaust sounding as though the log train was "coming like 60" down the track. The movie train carried two carloads of passengers, mostly extras who had been recruited around Chico. They were not aware of the discrepancy between speed and sound of a Shay and fearing an imminent rear-end collision, promptly started abandoning the train to its fate. The engineer of the movie train stopped as soon as the passengers started jumping off, and the log train finally came into sight around the curve. The Shay was making all of 10 miles per hour. The engineer (believed to have been Ed Myers on No. 102) stopped when he saw the mainline was blocked, and then backed up to allow the movie train to retreat to its spur.

The "Lafayette" and its crew had other problems during the twelve day stay at Chico Meadows. One unique problem occurred when the "Lafayette" became mired in mud at a road crossing, an occurrance somewhat more common with automobiles than steam locomotives! They got into the predicament while turning the "Lafayette" on the wye that had been built for the movie train's use. The wye sloped downhill and had a road crossing it to the movie property tents. Some snow had fallen and the crossing became coated with a layer of slippery mud. The crew backed the "Lafayette" down the wye, only to find the engine couldn't make it back uphill across the crossing. Each time they tried to go forward the engine slipped farther back, and then the tender trucks derailed. The engine suffered the ignominy of being pulled from the mud hole by a tow truck.

Diamond purchased two additional Shays in 1941 and placed them in service during the 1942 season. The first Shay was a small two-truck 45-ton engine owned by Diamond's Northwest Timber Division since 1936. It was assigned to the Albeni Falls, Idaho, sawmill as the mill switcher until replaced by a switching agreement with the Great Northern in March 1941. The Shay, shipped to Stirling City during the summer, became No. 105 on the roster. Too light for work on the mainline, it was assigned to the mill switching job where it displaced one of the Porter saddletankers.

Wallace Beery (left) and Robert Taylor who starred in "Stand Up and Fight." (Isabel Millington)

*An excellent account of the filming of "Stand Up and Fight" and "Wells Fargo" is contained in Mr. Sagle's article, "Lights! Camera! Action! Recollections of a Man who Stabled Iron Horses for Hollywood," in the April, 1962, issue of *Trains.*

The temporary Powellton camp was abandoned in late 1935 and replaced by the new Butte Meadows camp At left, newly purchased Crane No. 2 loads a two-man cabin at Powellton, while at right a train load of cabins leaves Powellton. (Beryl Orner collection)

Butte Meadows, officially Camp No. 1. It stood about a mile above the still-existing settlement of Butte Meadows. The logging railroad passed near the cookhouse at upper left. The large building in center was the community hall. (John Hamby photo, Edna Rouse Thelander collection)

The Butte Meadows camp cookhouse. The mainline of the logging railroad passed directly in front of it. (Dana Bailey)

Shay No. 104 derailed on the mainline not far out of Stirling City, and No. 102 – after pulling the train back to town – came back to assist. The time: the late 1930's. (H. A. Otis collection)

Extended logging operations out of Butte Meadows and the need to have an extra Shay large enough to work both the mainline and spurs prompted the decision to obtain another Shay in 1941. After several inquiries regarding engines for sale, Master Mechanic Al Myers and Trainmaster Stan Keesy went on a locomotive inspection tour of Oregon and Washington. After looking at a number of Shays for sale – including one almost identical to No. 102 – they chose a three-truck 90-ton Shay at Carlton, Oregon. It was owned by the Alaska Junk Company, which had obtained the assets of the bankrupt Flora Logging Company. Harold Beavers, who had operated the jammer since 1927, went to Salem, Oregon, on February 20, 1942, to be engine messenger on the Shay. The trip over the Southern Pacific took seven days, part of which time the temperature was seven degrees below zero. The Shay received an overhaul at Stirling City before going into service as No. 106, at the beginning of the 1942 season.

Beginning with the 1942 season, a pattern developed in the railroad operations that lasted until the line was abandoned. No. 106 was usually the woods switcher based at Butte Meadows. Willamette No. 104 was the mainline road engine working out of Stirling City to Butte Meadows, and No. 102 remained at Stirling City as the spare for both engines. No. 105 was the mill switcher, with its spare either No. 5 or 6. One saddletanker was kept at Chico as the Lumber Department switcher, while the extra saddletanker was spare for both Chico and Stirling City. No. 101, no longer needed as a spare after the arrival of the additional Shays, was scrapped at Stirling City in 1942.

An interesting sidelight to the pattern of operations starting in 1942 was that the locomotives had virtually the same engineers during the entire period. M. B. "Brad" Smith had No. 104, Harold Beavers was on No. 106, and Kimshew veteran Harry Marion ran No. 105. Each man was very partial to his engine.

The schedule called for the mainline run to leave Stirling City about 7:30 a.m. for an early afternoon arrival back in town if the crew had no trouble. The usual train was composed of 20 to 26 cars. If the road crew had a heavy train, the mill switcher served as helper engine pushing at the rear on the grade out of town cutting off at its crest near the town reservoir. The road crew had a long day, as they switched the loads to the mill after arrival back in town. If the road crew went all of the way to Butte Meadows, the practice was to leave the empties on the lower part of the balloon loop just below the camp. Then they would cut the engine off, go to the upper side of the balloon loop, couple onto the loads left by the woods crew and be out of town in minutes. They usually did not take water at Butte Meadows as there were enough tanks along the mainline.

Part of the time, in an effort to cut the road crew's day, the woods crew started taking the loads to Alder Creek Siding three miles towards Stirling City. There they also picked up the empties. This was abandoned after a trial, because the road engine had to back up when going out with the empties, and No. 104 tended to climb off the rails when backing up. A heavy load made it worse, as it would stiffen the gears and the big Willamette wasn't very flexible in the first place.

The crews had most of their troubles with car de-

Shay No. 102 crossing Deer Creek Highway Bridge with a trainload of logs bound for Stirling City. (John Hamby photo, Stanley Keesy collection)

railments instead of the Shays getting off the track. Sometimes only the rear wheel of a rear set of car trucks would jump the track. These were especially hard to see. One of the worse derailments occurred when one truck under a loaded car jumped the rails, just east of the Secret Creek Siding. Since it was night, the derailment was not seen or heard by the crew. The wheels rode the ties for over two miles, shearing the track bolts and angle bars. Near Swede Cut another pair of wheels jumped the track. One brakeman heard the noise this time and signaled the engineer, but before they came to a stop they had two more cars on the ground. The crew took No. 104 and the cars ahead of the derailment into Stirling City. The next morning they headed back to re-rail the cars. The woods crew started at approximately the same time from Butte Meadows to pull the tail end of the train back to the siding. The damaged track behind the train was discovered just in time to flag Harold Beavers on the 106 before he reached the torn up track.

When crews worked at night, they tied a lighted lantern to the last car. The crew thus could tell if its train was still on the track and in one piece, as Diamond didn't use cabooses. One night the woods crew was bringing a string of loaded cars off a spur above Butte Meadows. Looking back, they saw the lighted lantern bouncing among the trees some distance from the track. They stopped to investigate and discovered several cars on the rear of the train had jumped the track. The cars completely left the right-of-way, rolling into a nearby meadow.

The company on December 13, 1943 suffered the worst logging railroad wreck in the history of the Stirling City operations. It involved both cars and a locomotive at approximately 6 p.m. when darkness was beginning. The wreck was on the Jonesville Spur just above Butte Meadows. The practice of the woods crew was to leave the empties on Spur 45A, just above its junction with the Jonesville Spur, while they went up the Jonesville Spur to get the loads at the jammer.

The woods crew wasn't working that day, so the mainline crew with No. 104 was doing the switching. They had set ten empties on Spur 45A ahead of three loads. After leaving the loads from the jammer above

Three-photo sequence of the "Bull Creek Loop" on the mainline to Butte Meadows. (Top) A sharp 40-degree curve takes the track over Bull Creek. (Center) A Stirling City-bound train headed by No. 104 eases around the hill. (Bottom) No. 104 takes water at the Bull Creek tank as the brakeman checks the train. (John Hamby photos, top and bottom, Edna Rouse Thelander collection; center, Robert Lowry collection)

No. 106 pauses for water at the Bull Creek tank (left), while at right, No. 106 has just brought a trainload of logs to the siding one mile west of Butte Meadows camp, where the loads and empties were exchanged. (Left, Edna Rouse Thelander collection; right, Ernest Van Sant collection)

the switch on the mainline, they started back up the Jonesville Spur for the empties on Spur 45A, which were to be taken to the jammer. In the meantime, the air must have leaked off the empties, which rolled into the three loaded cars and started them. The whole string then rolled down the 3½ per cent grade into the Jonesville Spur, picking up momentum as they rolled down the spur towards the unsuspecting crew. It was dusk and the lack of warning was compounded by the runaway string of cars shooting around a curve only a hundred yards from the engine — which was backing up. The rear-end collision instantly killed Conductor William Flanagan and Brakeman Arthur Fierre, both standing on the tender steps. The terrific impact threw the engine's throttle wide open and the 70-ton engine was knocked more than 200 yards down the track. Despite the impact, the 104 stayed upright and on the rails. The engine's tank broke loose, twisted to one side, and was shoved into the cab against the backhead of the boiler. Both Engineer Brad Smith and Fireman Herman Fieth were penned in the wrecked cab, but the backhead of the boiler saved them from being crushed to death by the tank. Brad Smith remembers telling his fireman to blow the emergency signal on the whistle for the section gang before he lost consciousness. Both the thundering crash and the whistle signals were heard in Butte Meadows and before long the section gang and every available logger reached the wreckage. Rescue crews from Butte Meadows, Stirling City and Chico worked throughout the night to free the trapped enginemen and to locate the remainder of the crew. It took more than three hours to free the engineer and fireman from the wrecked cab and they did not reach Enloe Hospital in Chico until after midnight. Both recovered from their injuries.

After freeing the engine crew, the rescuers concentrated efforts on finding the conductor and brakeman. Members of the rescue crew that cold December night remember the wreck as a terrible mass of jumbled, wrecked flat cars. The three loaded cars were the first to hit the engine and the impact broke the binder chains holding the logs. The logs were scattered everywhere — even on top of the engine. Most of the flat cars were piled-up around and behind the engine, but a few were quite a distance from the track. The bodies of the conductor and brakeman were not found until morning under the mountain of wreckage, and were picked up in pieces.

The badly damaged locomotive was towed to Stirling City, presumably to be scrapped. The damage included the smashed fuel and water tank, wrecked cab, wrecked drive shaft, the rear set of trucks completely ruined and the frame was broken. Consideration was at first given to purchase of a replacement Shay, but second-hand Shays were virtually unobtainable during the war years. Master Mechanic Al Myers estimated it would cost $3,000 to repair No. 104 at the Stirling City shop. This was cheaper than an equivalent sized geared engine could be purchased. The decision was to repair No. 104, and after months of work at the shop, she returned to her log hauling duties during the 1944 season.

It was during the 1949 season that the last wreck of any consequence occurred on the logging railroad, and again it involved No. 104. She was on the point of 22 loads when she struck a broken rail on a curve between

No. 104 went over the bank near Bottle Creek in 1949 after hitting a broken rail. Above left is a close-up of No. 104. (Beryl Orner collection) At right, No. 106 approaches the end of the 22-car train to haul the cars back to Butte Meadows. (H. A. Otis)

Three loggers buck a 16-foot log off a 70-foot Ponderosa Pine at Willow Creek using an Atkins Hassler electric power saw. (Rognon Studio, D. I. Corp. collection)

mileposts 18 and 19 before Brad Smith could get stopped. The big Willamette jumped the track and headed over the bank, but luck was with the crew that day. The engine coupler twisted in the coupler of the first car and they held. That was all that kept the 70 ton locomotive and her crew from sliding 1,500 feet down the steep mountain-side into Bottle Creek. It required a crane and two Caterpillar tractors to bring the engine up.

Stirling City in 1950 witnessed a diesel invasion of the logging railroad, with testing of a General Electric 70-ton model demonstrator unit. It first relieved No. 105 as mill switcher and was well liked by Harry Marion. The diesel then made a trial run to Butte Meadows, but there it failed to live up to expectations. The builder's representative stated the unit would pull the same amount of tonnage as No. 104, Diamond's most powerful engine. The diesel tried 22 cars out of town (No. 104's tonnage rating) and couldn't make it. On its second try, it took 20 cars – the rating of No. 102 – without assistance. The 84-inch wheel center proved to be too rigid for some of the sharp 50-degree curves on the mainline and the oiled rail on these curves (used to help cars negotiate them) resulted in constant wheel slippage. The diesel also kinked the rails on the curves, making it rough riding for the Shays and requiring a month's work for the section gang. Only one trip was made with the diesel into Butte Meadows. The salesman left Stirling City without getting an order.

Despite the difficulties of the diesel on its trip over the mainline, the test was relatively successful. The diesel cut fuel costs in approximately half and hauled nearly the same tonnage. Another factor was the condition of Diamond's motive power, since only two of the four Shays were in good condition. No. 102 had seen very little service since 1946 and was virtually worn out. No. 104 suffered from worn gears and a cracked frame, which kept breaking despite repeated welding. (This was due to one of her many wrecks.) The important factor, her boiler, was still good. With repairs, the engine promised several more years of service. Nos. 105 and 106 were in excellent condition, with even more years of service than could be expected of No. 104.

Because of the test diesel's difficulty on the curves, Diamond planned to purchase two smaller units: probably General Electric's 44-ton model, and equip them for multiple unit operation. This, however, remained only a proposal and then a decision was made to abandon the entire railroad.

The abandonment decision was caused by a number of factors. The railroad had hauled less logs the last few years because the company cut in areas that had no railroad. Logs were trucked directly to the mill pond. Costs charged against the railroad steadily increased, mainly for paying a large section crew to keep the track in shape. Even with this expenditure the track remained in poor condition, causing many derailments. Part of this was due to the company never having sufficiently heavy rail for its operations. No. 104, particularly, was too heavy for the track.

Remains of the Bolt Creek trestle and grade near Butte Meadows camp in 1971. (Photo by the Author)

In 1953 the entire railroad and its equipment was sold as junk to the Commercial Construction Company of Dallas, Texas. Shay No. 106, with Brad Smith at the throttle, powered the scrap train taking up the rail. The first locomotive to be scrapped was No. 102, followed by No. 104. When the scrap train reached Stirling City, No. 106 followed her sister engines to the scrap pile.

No. 105 was not scrapped immediately, since light rail in the mill yard prevented Southern Pacific's locomotives from getting on many of the spurs to switch. The little Shay's days were numbered, however, as heavier rail was laid. Diamond's last Shay was scrapped during the summer of 1954. A half century of railroad logging at Stirling City ended.

The McGiffert loader in action in 1932. Harold Beavers (engineer) on the "jammer," with fireman Bill Andrews. Dana Bailey stands in front of the jammer; behind him is Wilbur "Shorty" Paul (top loader). Joe Walker (loader) holds the cable. (Harold Beavers collection)

Two pictures below show the jammer near Camp No. 2 in 1935. Engineer Harold Beavers works at a landing along the railroad line with his crew. Note the two types of log cars – old wood flats and skeleton flats – then in use. (Both photos by Elmer Marion)

No. 106 switching a single car at Butte Meadows camp in 1949. This view taken looking toward Butte Meadows. (Ted Wurm)

Same setup showing No. 106 backing up at Nigger Head Point just at the end of the passing track toward Stirling City. (Ted Wurm)

Nos. 104 and 105 on adjacent tracks in front of the Stirling City engine house in 1949. (Norman Holmes)

Mainstay of Diamond's logging equipment by the mid-1940's were nine D8 tractors with Hyster towing winches and arches. In these two views, taken November 8, 1945, snow was on the ground as (above) a D8 arrives at the landing. The operator of Crane No. 2 may use one of the logs to top off the load on the converted tank car. At left, another D8 pauses for its portrait before heading to the landing. (Both D. I. Corp. collection)

The first company-owned logging trucks – three Internationals and two Whites – came in 1939. Here are two Internationals with about 7,000 board feet on each truck. (D. I. Corp. collection)

Crane No. 1 unloading Truck No. 4, one of the Whites, at the Deer Creek Transfer on the Upper Cascade line about 1946. (Harold Beavers collection)

A truck load hits the pond from one of the Whites. The Internationals were Nos. 1-3; the Whites were Nos. 4 and 5. (D. I. Corp. collection)

A portion of the "junk yard" of railroad scrap at Stirling City during dismantling of the logging railroad in 1953. (Robert Lowry)

In another part of the "junk yard," Diamond's Crane No. 2 works for the scrappers. (Robert Lowry)

The Butte Meadows camp railroad yard in 1947, looking down the mainline toward Stirling City. (Ted Wurm)

CHAPTER TWELVE

Plant Railroads

DURING THE FIRST decade of its operation, the Barber (Chico) plant used a Butte County Railroad locomotive as its switch engine. Usually it was one of the 4-4-0 American Standards (Nos. 2 and 3), but occasionally a larger 2-8-0 would be pressed into service for a short time.

When Diamond assumed control over Sierra Lumber Company in 1907, there were two locomotives at that firm's Chico factory. The spare engine was the old Marysville Foundry Company engine built for the Red Bluff plant in 1877. Sierra replaced it in 1900 with a new Climax, and transferred it to Chico to replace the horses that were then being used to move cars. In 1906, it was again set out to pasture when Sierra purchased a new Porter 0-4-2T saddletanker to switch the Chico plant. The new Porter was No. 6 on Sierra's roster, while the old Marysville Foundry Co. engine is believed to have been No. 5.

Diamond took an appreciative look at the year-old Porter and moved it to the Barber plant to replace the switch engine supplied by the Butte County Railroad. No. 6 spent less than a year at the Barber factory, however, because Diamond decided to transfer her to Stirling City and went back once again to using the Butte County Railroad''s engines to work the Barber facility.

Almost immediately after Diamond purchased Sierra, work started on dismantling the machinery in Sierra's Chico Division factory and moving the piled lumber at the factory to Diamond's plant at Barber. The Barber switch engine and its crew had enough work to keep them busy within the plant without also having to work the 9th Street spur to the old Sierra plant. Another crew was assigned to this job. This second crew included engineer Charlie Marion and his son, Elmer, who were ordered to go over to the Sierra factory where they would find old No. 5.

The Marion father and son team, which formerly worked on Sierra's meter gauge logging railroad in Chico Canyon, had never seen the five-spot before. They were a little shocked when they saw it, their first reaction being that it looked like a "home-made" affair. They were even less impressed with the engine after they fired it up and tried to switch with it. It was an extremely poor performer to say the least. They heartily cussed their assignment on it. By that time they were hearing stories about the engine's poor performance on 9th Street before Sierra replaced it with the new Porter. The stories held that the crew had to stop at nearly every intersection to blow up steam before the engine could creep along again with only one loaded car. It was little wonder the old five-spot was such a poor performer. As Charlie Marion examined her, his experienced eye found she was in poor mechanical condition. Then the crew found several pounds of old nails in the bottom of the firebox, which didn't help the engine's steaming qualities. (The nails had been accumulated in the past from using old lumber as fuel.) They received permission to run the engine over to the Engineering Department at Barber for an overhaul. When they got it back, they had no further complaints about its operation.

The Marion father and son crew worked the old five-spot on the 9th Street spur for almost a year, hauling carloads of lumber and machinery out of the old Sierra plant. They also hauled in loads of veneering (plywood) machinery for the plant's conversion. The little woodburner surprised them with its power after its overhaul. On one trip from the plant they found four carloads of gravel blocking the spur between Main and Wall streets, where the Southern Pacific switch engine had spotted them for unloading

Porter-built No. 103 shown during switching chores in 1937. Engineer Charlie Marion poses at right with two flat cars loaded with box shook. (W. C. Whittaker)

gravel needed for street paving. The five-spot took all four loaded cars back to the plant in one trip, later returning them to where they had been spotted after completing the trip to the interchange track. Another time they hauled seven carloads of lumber to the interchange track in one trip. Once they switched for the Southern Pacific. This occurred one morning when they arrived at the interchange track to find their empty cars were mixed with a long string of cars on a siding and the Southern Pacific's switcher unavailable. After waiting more than an hour for the switch engine to return, they were given permission to do the switching. They had a relatively short time to finish their work between arrival of trains. To save switching moves they coupled onto as many cars as the engine could move. During one switching move they pushed 20 empty cars up the siding for about half a block. Their work took them from the 9th Street interchange up the mainline almost to the 1st Street crossing. During this move they passed the depot at 5th Street, where a crowd was beginning to gather for the arrival of a passenger train. Elmer Marion laughingly recalled to the author that several Southern Pacific employees were disclaiming to the people at the depot that there was any company connection with the odd looking woodburner and were carefully keeping their distance! By the time the passenger train arrived, Charlie Marion had his little woodburner chuffing back up 9th Street with its empty cars.

They had troubles, too. On one trip they ran out of both wood and water. Another time they started for the interchange track after not having made a trip for a week. The engine jumped the track in the middle of the intersection at 9th and Main streets, but Charlie Marion kept the throttle open until he cleared the intersection for the passage of traffic. It took them an hour to rerail the engine, using slabs of wood from the wood box.

The 9th Street spur was relaid with heavier steel early in 1908, enabling Diamond to haul logs directly from Stirling City to their new veneer plant. This facility went into production in April, but the five-spot and its crew had ended their work on the line by that time. With the plant going into production, there wasn't as much switching as while Sierra's factory was being closed, and Diamond decided the Barber switch engine could handle the veneer plant switching along with the switching at Barber. The Marion father and son crew fired up the old five-spot for the last time, pulled a few cars to the 9th Street interchange, and then ran a short distance down the Southern Pacific line to the Barber plant. At Barber, Charlie Marion was told to park the five-spot on the end of the spur next to the Engineering Department building and kill the fire. He misjudged the end of the spur, and the five-spot came to a stop in the dirt off the end of the track. They killed the fire and that was the last time they saw the little woodburner. The Marions transferred to the logging railroad out of Stirling City, and when Charlie Marion came back to Barber in 1916 to be the engineer on the switcher, the old five-spot was gone.

The Butte County Railroad continued to furnish the switcher for the Barber plant until November, 1915, when Diamond's lease expired and the railroad was transferred to the Southern Pacific for operation. Just before the lease expired, Diamond moved the Porter 0-4-2T saddletanker from Stirling City to Barber. This little saddletanker, formerly No. 6 on both the Sierra and Diamond rosters, had been renumbered to No. 103 in 1911 to avoid conflicting with the number assigned to a meter gauge Porter saddletanker purchased that year. Meter gauge operations were ending at Stirling City in 1915 and Diamond had surplus motive power for the short standard gauge logging railroad that was planned for the next season.

There was concurrently a great curtailment of operations at the Barber plant, leaving the match factory as the only portion of the Barber plant in production between 1916 and 1920. At this time the plant switch engine and its crew went under the jurisdiction of the

Match Department. Even after the Lumber Department was reactivated at Barber in 1920, No. 103 remained under the supervision of the Match Department but handled the switching for the entire plant.

No. 103 operated with a two-man crew, the brakeman being furnished by either the Match Department or the Lumber Department – depending upon whose cars were being switched. Charlie Marion was both engineer and fireman, as No. 103 was a "one man engine." This was possible due to the little saddletanker being an oil burner; the engineer did the firing by means of a rod extending from the firing valves on the left side of the cab.

The saddletanker was housed in a small, one-stall corrugated iron shed behind the match factory when not in use. Here Charlie Marion performed light repairs on her, but if the locomotive needed heavier repairs than he could handle, No. 5 or No. 6 was deadheaded from Stirling City in a Chico-bound freight. No. 103 was then deadheaded to Stirling City, where the necessary repair work was done at Diamond's enginehouse. The Stirling City facility had more shop machinery for locomotive repair than what was available at the Chico plant. When the repair work was completed, No. 103 was deadheaded to Chico and the relief engine returned to Stirling City.

The roster of equipment at the Chico plant, besides No. 103, consisted of several ancient flat cars. At least one of these relics was off the old Butte County Railroad and still carried fading lettering for this company into the 1940s. Due to their age the flat cars were confined to use within the plant and could not be used in interchange service.

The advent of the World War II in 1941 greatly increased Diamond's business, resulting in a much heavier switching load for the plant railroad. Both No. 5 and No. 6 were brought from Stirling City to help with the increased switching. (This was possible for the first time due to Diamond purchasing two secondhand Shays in 1942, one of which was assigned to the sawmill as switcher.)

A few log flats were also brought from Stirling City and three rather unusual cars were purchased. The Sacramento Northern in 1939 abandoned through electric interurban service from Oakland to Chico. Subsequently most of the cars were brought to the electric railroad's Mulberry Shops in Chico for scrapping. By late 1941, the scrapping of the historic interurban cars was at its peak only a few blocks away from Diamond's plant. Apparently one of Diamond's local officials saw an opportunity for the company to acquire used equipment at good price. The interurbans could then be cut down to flat cars for plant use. Diamond obtained

Flat car No. 3 was rebuilt into a homemade box car to haul box shook prior to this 1938 photo. (Mavin Maynard)

Diamond purchased three retired Sacramento Northern interurban cars in 1940 for use in hauling box shook. Behind No. 21 (bottom) is the former Engineering Dept. foundry and machine shop. (Vernon Sappers)

The Chico plant diesel spots a car adjacent to the match factory in November, 1963.

Earlier the same day, the diesel moves two cars across West 16th Street.

A secondhand General Electric diesel, similar to the Chico unit, was purchased for switching the Red Bluff Integrated Forest Products factory. (Three photos by the Author)

three interurban trailers from Sacramento Northern. These were Nos. 1202, 1203 and 1209, which were numbered 20, 21 and 22, respectively.

Each new acquisition had seen many years of service. Nos. 1202 and 1203 were built in the 1880s as passenger coaches for the Central Pacific Railway. For a number of years they served local suburban lines operated by the Southern Pacific in the East Bay area around Oakland. In the early 1913, the Southern Pacific sold six of these local suburban coaches to the Oakland, Antioch & Eastern Railway (a predecessor of the Sacramento Northern). The O.A. & E. rebuilt the cars, numbered them 1201-1206, and placed them in service during April, 1913, as trailers in their local East Bay service. These cars were very long for interurban usage. No. 1202 was 60 feet long, while No. 1203 was 62 feet, 8 inches in length.

Car No. 1209 was also historic in its own right. It was among a group of three cars which were built by the Holman Car Company of San Francisco in 1906 for the ill-fated Ocean Shore Railroad. In 1916 the three cars were sold to the O.A. & E., which numbered them 1208-1210. No. 1209 was shorter than the other two cars acquired by Diamond. It measured 52 feet, 1 inch in length.

After Diamond purchased the cars, the work of cutting them down to flat cars was deferred. Two large openings were cut on each side to enable the cars to be used for hauling box shook to the warehouse across West 16th Street from the millwork plant. Sometime during or just after the war, all three interurban trailers were cut down to flat cars. One was used for only a short time, until it was loaded excessively for its light wooden frame, which broke under the strain.

During the summer months the railroad operated on two shifts, as did the other departments of the plant. The day shift came on the engine at 6 a.m. to switch the cars before the day shift arrived in the plant. The railroad's day shift lasted until 2 p.m., when the night crew came on duty and worked the engine until 10 p.m. They didn't switch all of the time they were on duty. They waited between moves, but they were on call during their entire shift. The crew members even ate their lunch aboard the engine. Between ten and twenty cars a day were switched to the green chain, and the box factory and Millwork Department each accounted for four or five cars daily. At the Match Department No. 103 was used only on one shift throughout the year. The match factory accounted for approximately four or five cars of outgoing shipments daily with a lesser number of inbound cars. During the winter months the switching was lighter for the Millwork Department switcher, as the Stirling City mill was

No. 103 met her end on June 3, 1952. A portion of her remains is shown here on the scrap dealer's truck ready to depart from the plant. (Robert Lowry)

closed. Only one shift, from 8 a.m. to 5 p.m., was needed.

Nos. 5 and 6, more powerful than No. 103, could pull approximately fifteen loaded cars at one time. The balloon loop, originally constructed in 1904 to turn entire trains, was still available and was used for some of the switching moves and occasionally for car storage. The balloon loop was the scene of trouble one day when the crew of No. 6 decided to use it for making a switching move with one of the former Sacramento Northern interurban cars partially loaded with short lengths of lumber. The length of the old interurbans caused the crews problems on some spurs, particularly on the curves. The cars were hard to handle when they were loaded and due to their light wooden frames, could only be loaded directly over the trucks on each end.

On this particular day, No. 6 and crew started around the balloon loop with one interurban car in tow, but came to grief when their balky interurban derailed. The derailment on the sharply curved track bound the coupler between the car and the engine tight; the crew couldn't uncouple the engine. Finally after much work, sweat and cursing, they got the engine uncoupled and just in time. The boiler was nearly dry with no water left in the tank. The crew, to avoid the possibility of a boiler explosion, killed the fire. Then they went to the engine shed to fire up No. 5, which towed her sister engine back to the engine shed. Later in the day they rerailed the interurban. Operations returned to normal.

A General Electric 150-horsepower diesel replaced all three saddletankers in 1948. The diesel was only a few months old in this 1949 view. (Norman Holmes)

On June 1, 1946, veteran engineer Charles Marion retired; within a year his engine, No. 103, also was retired. The Millwork Department crew then started switching for the Match Department. Most of the time No. 6 was used by the crew with No. 5 serving as the spare. A General Electric 150-horsepower diesel was purchased in 1948 and both steamers were put out to pasture. Both saddletankers were stored at the engine shed: one was parked inside and the other outside. The diesel never called the engine shed "home" and has always been parked on Track 10 inside the Lumber Department warehouse. All three steamers remained in storage until 1952 when Nos. 5 and 103 were sold for scrap to Pollak Steel Supply. No. 6 lasted one more year until being sold; also for scrap, to the same company. At the same time the one stall engine shed was torn down.

The amount of switching for the plant railroad gradually declined over the next few years, and then dropped sharply with the opening of the Red Bluff Intregrated Forests Products plant in October, 1958, and the closing of the Stirling City sawmill the previous January. A similar General Electric 165 horsepower diesel, purchased for the Red Bluff plant, was used for several months at the Chico plant while its switcher was overhauled before being moved to the Red Bluff plant.

The schedule in the 1970s called for the Chico plant to be switched early in the morning and again early in the afternoon, when cars were to be moved. Rail shipments in the 1970s amounted to between five and ten cars each week, mostly from the Millwork Department. Both the diesel and its engineer are under the Yard Department. A switchman is furnished by the Millwork Department or the Match Department, depending upon whose cars are being switched at the time. Maintenance of the diesel is handled by the company's garage mechanic.

At the Red Bluff Integrated Forest Products Center, the switch engine is operated by a two-man crew under the jurisdiction of the Central Services Department. Rail shipments are much heavier out of the Red Bluff plant and the crew is kept busy from 7 a.m. until after 5 p.m., except for short periods between moves. There are six spurs within the plant with the total length about two miles.*

*Additional information on the operation of the plant railroad of the Red Bluff Integrated Forest Products Center is contained in the author's article "Diamond National Corporation" *Pacific News* (Issue 34, June 1964), p. 3-6.

Rosters

BUTTE COUNTY RAILROAD COMPANY
Roster

Locomotives

No.	Type	Builder	No.	Date	Cylinders	Drivers	Boiler Pressure	Tractive Effort	Wt.
1	2-8-0	Baldwin	20192	1902	20x28	50″	190	36180	180,000
2	4-4-0	Baldwin	9711	1888	18x24	61″	125	12668	96,000
3	4-4-0	Baldwin	8404*	1887	18x26	59″			102,000
4	2-8-0	Schenectady	29704	1904	22x26	51″	180	37750	170,000

NOTES

1. Built for Columbia Southern Railway, No. 3. Sold to W. D. Hopkins Company (Portland, Oregon) account too heavy. Sold 2-1903 to Butte County Railroad for $14,750. To Southern Pacific 12-1-1915, renumbered C. P. (Second) 2503 in 1916, Class C-12. Scrapped at Sacramento, 9-25-1936.
2. Built for Elgin, Joliet & Eastern, No. 21. Purchased by Butte County Railroad in mid-1903 from F. M. Hicks Company (Chicago) for $6,000. To Southern Pacific 12-1-1915, scrapped at Sacramento 11-1916 without being renumbered C. P. (2nd) 1304, Class E-33.
3. *Tentative identification of this locomotive is Baldwin c/n 8404, built as Union Pacific No. 744. As built the specifications were 62" drivers, 18 x 26" cylinders, tractive force 18478, weight 99000. Locomotive evidently later rebuilt while still under U. P. ownership. Vacated from U. P. roster 5-1902, sold with a group of other U. P. 4-4-0's to F. M. Hicks Company. Butte County Railroad No. 3 definitely purchased from F. M. Hicks Company sometime during 1903, on roster by 11-27-1903. To Southern Pacific 12-1-1915; apparently had been out of service for sometime and was towed to Sacramento in partially stripped condition. Scrapped 6-12-1916 without being renumbered.
4. Only locomotive built new for Butte County Railroad, delivered to Barber on 6-1-1904. To Southern Pacific 12-1-1915, renumbered C. P. (Second) 2502 in 3-1916, Class C-11. Scrapped at Sacramento, 10-19-1935.

Equipment: Freight

201-230	flat cars	35′7″	80,000 lbs. cap.	Illinois Steel Co.
231-260	flat cars	40′8″	80,000 lbs. cap.	(Lebanon, Pa.)
261	flat car	39′11″	80,000 lbs. cap.	built 10
501-505	box cars	34′2″	60,000 lbs. cap.	American Car & Foundry
506	box car	34′2″	50,000 lbs. cap.	(St. Louis)
?	caboose			

In 1912 B.C. car Nos. 223, 237, 240, 241, 243, 244, 246, 251, 253, and 255 were equipped with temporary fuel racks.

Equipment: Passenger

1001	Combo	Purchased from Heiley & Co. secondhand
?	Baggage car/smoker	Pullman, 1904
?	Coach	Pullman, 1904
?	Observation coach	Pullman, 1904

Builder's photo of Butte County Railroad No. 4 (Guy Dunscomb collection)

No. 1, probably during construction of the railroad. (John Nopel collection)

Former Butte County Railroad No. 4 as Southern Pacific No. 2502 (2nd) stored at Sacramento in 1934. (Guy Dunscomb)

No. 2 on a work train at Stirling City. (Mrs. E. L. Myers collection)

SIERRA LUMBER COMPANY
Locomotive Roster

No. and Name	Type	Cylinders	Drivers	Weight	Builder	No.	Date
Red Bluff Division (Lyonsville) meter gauge							
1 "Peter Dean"	0-4-OT Rebuilt to an 0-6-OT about 1890. Transferred to Chico Division, West Branch Mill, in 1901, renumbered to No. 2				Prescott, Scott & Co.	20	1881
2 "Antelope"	0-6-OT	9x14	28"	25,000	Porter	564	4/1883
3 "Uncle Sam"	0-6-OT	10x14			Porter	1669	6/1896
Chico Division (West Branch) meter gauge							
1	0-6-OT	10x14			Porter	2032	8/1899
2 "Peter Dean"	0-6-OT Formerly No. 1 at Lyonsville.	10x12	30"		Prescott, Scott & Co.	20	1881
Red Bluff Division (Red Bluff) standard gauge							
? "Yellow Hammer"?	0-4-OT Original owner reported as Swain & Co. Ex-Sierra Flume and Lumber Co.; to Sierra Lumber Co. in 1878. Transferred to Chico in 1901.				Marysville Foundry Co.		1877
260	2T Climax			34,000	Climax	260	1901
Chico Division (Chico) standard gauge							
5? "Yellow Hammer"?	0-4-OT Transferred from Red Bluff in 1901.				Marysville Foundry Co.		1877
6	0-4-2T	12x16	33"	40,000	Porter	3574	1906

Meter gauge 39.36 inches.

All locomotives acquired by Diamond in 3/1907.

Red Bluff Division No. 2, the "Antelope," near Lyonsville before 1900. (D. I. Corp. collection)

Chico Division No. 1 near West Branch in Big Chico Creek Canyon. Charlie Marion in cab; Harry Marion in gangway.

Sierra Lumber Company Climax No. 260 at Red Bluff in 1901. (Both D. I. Corp. collection)

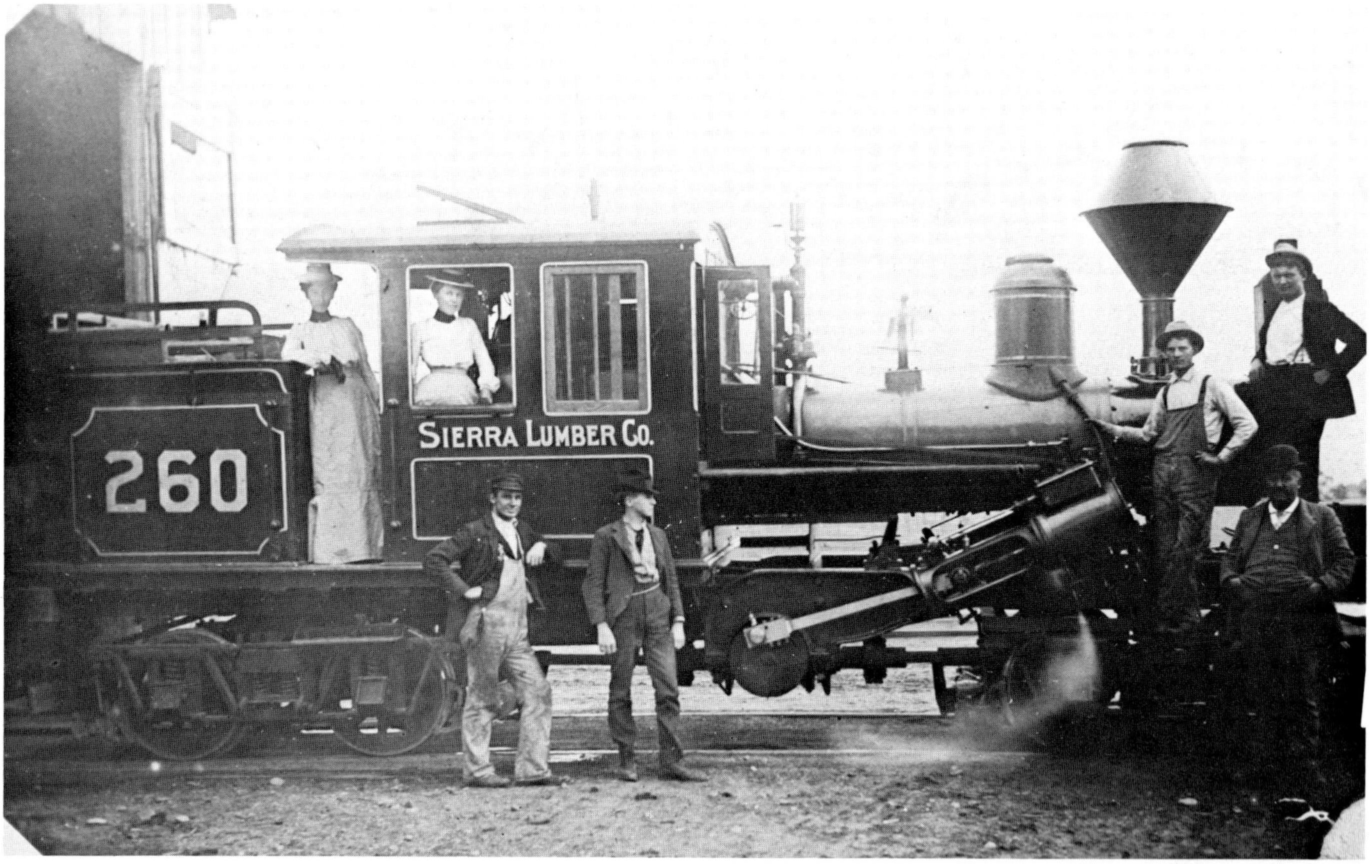

Above, No. 6 posed at Chico for this early day photo. (D. I. Corp. collection) Below, standard gauge switcher (Yellowhammer?) at Red Bluff in the 1880's.

DIAMOND MATCH COMPANY
Locomotive Roster (California Division)

No.	Type	Builder	No.	Date	Cylinders	Drivers	Boiler Pressure	Tractive Effort	Wt.
1*	0-6-OT	Porter	2032	8/1899	10x14				
2*	0-6-OT	Porter	564	4/1883	9x14				
3*	0-6-OT	Porter	1669	6/1896	10x14				
4*	0-6-OT	Prescott, Scott & Co.	20	1881	10x12	30″			
5 (1st)?	0-4-0	Marysville Foundry Co.		1877					
5 (2nd)**	0-6-0T	Porter	4809	2/1911	12x18	34″	180	11660	55,000
6 (1st)	0-4-2T	See No. 103 below							
6 (2nd)**	0-6-0T	Porter	4810	2/1911	12x18	34″	180	11660	55,000
101	2T Shay	Lima	860	1904	12x12	36″	180	20670	110,000
102	3T Shay	Lima	1748	1906	12x15	36″	180	25870	130,000
103	O-4-2T	Porter	3574	1906	12x16	33″	150	6180	40,000
104	3T Shay	Willamette	27	1927	12x15	36″	200	36000	143,180
105	2T Shay	Lima	3245	12/1923	10x10	29″	180	14320	81,000
106	3T Shay	Lima	3009	1918	12x15	36″	200	30350	146,000
260	2T Climax	Climax	260	1901					34,000
Diesel units									
796	B	General Electric	27811	8/1944	150 horsepower		25 ton		
870	B	General Electric	29365	4/1948	150 horsepower		25 ton		

*Meter gauge (39.36 inches)

**Originally meter gauge, rebuilt to standard gauge in 1915/1916.

NOTES

1. Sierra Lumber Co. 2nd No. 1 (Chico Division). Acquired by Diamond in 3-1907. Sold to Nine Lumber Company, Bray, Calif. in 1917.
2. Sierra Lumber Co. No. 2 ("Antelope"). Acquired by Diamond in 3-1907. Scrapped or sold after 1915.
3. Sierra Lumber Co. No. 3 ("Uncle Sam"). Acquired by Diamond in 3-1907. Sold to Nine Lumber Company in 1917.
4. Sierra Lumber Co. 1st No. 1, 2nd No. 2 ("Peter Dean"). Acquired by Diamond 3-1907; scrapped or sold after 1915.

   According to a retired employee, Nos. 1 and 3 were sold to the Nine Lumber Company. Initially the company bought one of the locomotives but came back for the second a few weeks later after an inexperienced engineer wrecked the first engine on their logging railroad. The same source also said that Nos. 2 and 4 were sold to F. L. Bottsford, San Francisco railroad equipment dealer, for scrap.

5. (1st)? It is not definitely known if there was a No. 5 but it is believed to have been Sierra's old Red Bluff switcher built by the Marysville Foundry Company. Acquired by Diamond 3-1907, used on the 9th Street spur in Chico 1907-08. Believed to have been scrapped by the Engineering Department at Barber after 1908.

5. (2nd) Built new for Diamond as a meter gauge locomotive designed for later conversion to standard gauge. Converted to standard gauge 1915-16. Scrapped at Chico by Pollak Steel Supply in 1952.

6. (1st) See No. 103 below.

6. (2nd) Same as 2nd No. 5. Scrapped at Chico by Pollak Steel Supply in 1953.

101. Purchased new by Diamond. Scrapped by Diamond at Stirling City in 1942.

102. Purchased new by Diamond. Scrapped by Commercial Construction Company at Stirling City in 1953.

103. Sierra Lumber Company No. 6, purchased for service on the 9th Street spur in Chico. Acquired by Diamond in 3-1907, 1st No. 6, used at Barber and then transferred to Stirling City. Renumbered 103 when 2nd No. 6 was purchased in 1911. Transferred back to Barber in 1915. Scrapped at Chico by Pollak Steel Supply in 1952.

104. Purchased new by Diamond, delivery point Stirling City. Scrapped by Commercial Construction Company at Stirling City in 1953.

105. Dalkena Lumber Co. No. 1, Dalkena, Wash.; acquired by Diamond Match Company, Northwest Timber Division, in 1936, No. 5. Assigned to Albeni Falls, Idaho sawmill as plant switcher 1936-1942. Transferred to California Division in 1942, assigned to Stirling City sawmill as plant switcher. Scrapped at Stirling City, 1954.

106. Siems, Carey — H. S. Kerbaugh Corporation No. 701, Joyce, Wash.; Anderson & Middleton Lumber Company, Hoquiam, Wash.; Willamette Iron & Steel Company, Portland, Oregon (dealer); Snow Creek Logging Company, Blyn, Wash.; Webb Logging & Timber Co., Brinnon, Wash.; Flora Logging Co. No. 4, Carlton, Oregon; Alaska Junk Co., 1940 (dealer); acquired by Diamond 1-1942. Scrapped at Stirling City by Commercial Construction Company in 1953.

260. Sierra Lumber Company No. 260, used at Red Bluff. Acquired by Diamond 3-1907, used at Red Bluff until 1911 or 1912. Disposition unknown.

796. Naval Ammunition Depot No. 2, Westminster, Calif.; U.S. Navy No. 65-00320; Pan American Engineering Company, Dallas, Texas (dealer): purchased by Diamond Match Company in 1956, No. 796, delivery point Chico. Assigned to Chico plant 1956-57, transferred to Red Bluff Integrated Forest Products plant in 1958. Diamond Gardner Corporation No. 796 (1957-58); Diamond National Corporation No. 796 (1959-71); Diamond International Corporation No. 796 (1972-); in service at Red Bluff plant.

870. Purchased new by Diamond Match Company, assigned to Chico plant upon delivery. Diamond Gardner Corporation No. 870 (1957-59); Diamond National Corporation No. 870 (1959-71). Diamond International Corporation No. 870 (1972-); in service at Chico plant.

Meter gauge No. 3 at Transfer Camp. (Elmer Marion photo, D. I. Corp. collection)

No. 6 as a meter gauge locomotive. The late Frank Compton sits on the pilot. (Frank Compton collection from D. I. Corp. files)

No. 4 at Ramsey Bar. (Bert Ward collection)

Standard gauge No. 5 at Chico in 1940. (Robert Hanft)

Standard gauge No. 6 switching at Stirling City in 1939. (Douglas Richter)

First No. 6 at the Butte County Railroad's Barber roundhouse in 1907. (D. I. Corp. collection)

Renumbered No. 103 in 1911, the same engine is shown here in 1946 next to the one-stall enginehouse at the Chico plant. (Robert Hanft)

Climax No. 260 at Red Bluff, sometime between 1907 and 1912. (Henry Schafer collection)

No. 101 as built. (Mrs. E. L. Myers collection)

No. 101 shown in 1941 at Stirling City. (Bert Ward)

No. 102 shortly after delivery. (R. S. Polkinghorn collection)

No. 102 at Butte Meadows in May, 1948. (Douglas Richter)

Builder's photo of No. 104 at the Willamette Iron & Steel plant at Portland. (John Labbe collection)

No. 104 at Stirling City in November, 1939. (Douglas Richter)

Willamette No. 104 at Stirling City enginehouse in 1949. Harold Beavers in the gangway; Brad Smith leans out of the cab window. (Ted Wurm)

Northwest Timber Division No. 5 (above) became No. 105 after transfer to California. This picture taken in 1948. (Bob Larsen collection)

No. 105 (left) shown in 1949. (Ted Wurm)

Diamond Match Co., Northwest Timber Division No. 5 at Albeni Falls, Idaho, plant in 1936. (Diamond Good Fellowship Bulletin)

Left side of No. 106, Stirling City, July 8, 1953, shortly before the Shay was scrapped. (Robert Lowry)

No. 106 as Flora Logging Co. No. 4 at Carlton, Oregon, on October 10, 1940. (Gerald Best)

Lima builder's photo of No. 106 in 1918. (P. E. Percy collection)

No. 796, Red Bluff Integrated Forest Products Plant switcher at Red Bluff in 1973. (Bill Shippen)

No. 870, the Chico Plant switcher, in November, 1963. (Kent Stephens)

Data for demonstrator diesel tested at Stirling City in November, 1950:

General Electric Co. No. 1: B-B G.E. 30448 7-1950 660 hp 70 ton. Demonstrated on several railroads in the western part of the United States, including the Apache Railway and the Sierra Railway. From Sierra it went to Diamond Match Co. at Stirling City. Following its trial on the logging railroad, it may have demonstrated on one or two other companies. Sold to Southern Pacific, repainted and renumbered, No. 5114, at Klamath Falls, Oregon, on January 2, 1951. Vacated from roster during 1967-68 fiscal year, sold to National Metal & Steel Corporation, Terminal Island (Los Angeles) as their No. 511 on October 27, 1967. Used as switcher until sold to Chrome Chrankshaft Company (Dealer), Los Angeles, during July 1972. Sold to Asarco Mexicana S.A., No. 1956, during August 1972. After repainting it was shipped to the Mexican company but was stored enroute at Southern Pacific El Paso (Texas) roundhouse pending decision by Asarco Mexicana S.A. if the unit was needed. Chrome Chrankshaft Co. was advised by the Mexican company that the unit was for sale; they put officials of Modesto & Empire Traction Co. (Modesto, Calif.), who were looking for another 70 ton G.E. unit, in touch with the Mexican company. Mr. Frank Sherman, Superintendent of the M&ET, went to El Paso to inspect the unit, and it was purchased by M&ET from Asarco Mexicana S.A. during December 1972. Renumbered M&ET 606 during February 1973, currently in sevice.

The demonstrator diesel shown as Modesto & Empire Traction No. 606 in March, 1975. (Tom Gray)

DIAMOND MATCH COMPANY
Equipment

Unfortunately all records concerning the equipment of the logging railroad have been destroyed, even the records covering the last years of the railroad's operations. The information below has been pieced together from other sources, including Diamond International Corp archives, notes, photographs, mention in newspapers and interviews with retired employees. Especially helpful were Stanley Keesy, the late Harold Beavers and Dana Bailey, who provided the data for the 1916-1953 period.

1904-1915:

Initial purchase of logging equipment was in 1905 when 40 flat cars were obtained from Western Steel & Foundry at $555.00 each. Also purchased were 10 Roger ballast cars at $425.00 each. The company then exchanged 6 of the ballast cars for 8 "dump cars" from the Butte County Railroad.

Equipment listed in various editions of the Official Railway Equipment Register (Standard gauge):

Nos.	Type	Length	Width	Capacity	
601-640 641- ?*	flat cars	40′ 0″	9′	80,000 lbs.	Added 1906?
701-731	flat cars	39′ 11″		80,000 lbs.	Added 1904-

Nos. 641—are not listed in the Official Railway Equipment Register but several cars numbered higher than Nos. 640 are shown in photos. Nos. 601-640 are not shown earlier than 4-1906 edition.

Kimshew operations (standard gauge):

One self-propelled log unloading car was used at the Transfer, which had probably been built by the Engineering Department. This car may have been the same one used earlier at the log pond at Stirling City.

Flat car No. 722 was rebuilt to an oil car to haul fuel oil to the Transfer, where the oil was transferred to meter gauge cars.

Kimshew operations (meter gauge):

Inherited Sierra meter gauge skeleton log bunks were augmented by 70 similar cars built by the Engineering Department for the 1910 season.

Four fuel oil cars were built from four old Sierra boilers and mounted on cars.

One heavy duty flat was used for moving donkeys, etc.

There is mention of speeders, both standard and meter gauge, but there is no record of how many the company owned.

1916-1953. Logging flat cars:

The railroad had a total of 100 to 110 cars at any given time. Until the 1930's these were divided into two types — flat cars and skeletons. The company originally purchased the flat cars, but by the 1920's most of these were being built in the Stirling City shop. All of the skeletons were built by the shop. Both types were largely wooden construction.

During the period from 1930 to 1953, the company added other types of cars and a definite numbering system was developed by Stanley Keesy.

Equipment Roster (1930-1953)

Series	*Notes*
1-37*	"Old" cars acquired or built before the 1930's. Largely wooden construction, mixture of flat cars and skeletons. Some cars newer than 1930, which were built or rebuilt to replace those destroyed in wrecks. The new car would assume the same number of the wrecked car.
40-70*	Steel flat cars rebuilt from retired oil tank cars. Starting in the mid-1930's, the company purchased these cars from F. L. Bottsford, a San Francisco dealer in used railroad equipment. A few cars were also purchased from another dealer, A. D. Schader. Bottsford purchased retired oil tank cars from both railroad and oil company car fleets. Once the cars arrived at his yard, the scrap crew removed the oil tanks and the cut-down car was subsequently sold to Diamond. At Stirling City, Diamond's shop installed log bunks, stenciled a number on the cars, and they were ready for service. The steel cars steadily replaced the wooden flat cars and skeletons as the latter wore out or were wrecked. The replacement program, however, was never intended to be complete. The main reason for this was the weight of the former tank cars. They had a steel underframe with steel side sills, resulting in a significant increase in weight over the older wooden cars. If they had a train of most or all steel cars, it was necessary to cut the train's length by three or four cars. Another disadvantage was that some portions of the car's underframe was hollow, making them extremely vulnerable to wreck damage to their frames. Once the frame was damaged the car usually had to be retired. In general, however, the steel cars were able to withstand much greater stress than could the wooden cars. Despite the increased weight of the cars, they were liked by the engineers due to their ten inch brake cylinder. All of the other cars had eight inch brake cylinders. An interesting sidelight is that several of the steel cars may still be in existence today. A train wreck near Swede Cut put five or six cars over the bank. Two of them slid so far towards Bull Creek that Diamond never retrieved them.
80-90*	Sixteen Russell cars purchased secondhand in Vancouver during 1944-45. They were too long for the sharp cures and had to be shortened by the shop.
100-105*	Eight retired Roma Wine refrigerator cars were purchased from F. L. Bottsford. Diamond converted six to flat cars, selling the bodies to farmers, etc., for storage sheds, but keeping several of the bodies for their own use as sheds.
120	Believed to be the highest numbered log car. This was a Russell car from the 80-90 series that was involved in the Swede Cut wreck that put several of the "steelies" over the bank. Rebuilt, it received its "maverick" high number after rebuilding instead of its former number.
297, ?	Two Roma Wine refrigerator cars were not rebuilt and were retained in their original form. Neither car was repainted or renumbered. Initially they were used part of each day as "cabooses" to take the crew out of the Butte Meadows logging camp in the morning and return in the evening. The men especially appreciated this during the cold months. The interior was fitted up with benches and a wood burning stove, the doors being left open for ventilation. These cars were the only cabooses ever owned by Diamond. After serving for a time as "cabooses," one car was used as the supply car for the Butte Meadows camp, spending most of its time in the camp except for an occasional roundtrip to Stirling City. The other car was used as a mobile supply car for "cat" parts. During the logging season it was stationed on a spur near where the Caterpillar tractors were working in logging. At the end of the season, the car was brought to Stirling City to spend the winter on the speeder spur next to the "cat shop." Eventually the cat parts car was removed from its trucks and set on the ground for use as a storage shed. The Butte Meadows supply car, No. 297, survived until the railroad was abandoned and was sold by the scrapping company to the operator of a combination grocery and bar at Butte Meadows. The de-trucked car was placed on the ground at the rear of the building, located at the junction of Humboldt Road and Bull Hill Road. At this writing, the grocery store and bar is known as the Bambi Inn and Roma Wine (C.L.D.X.) No. 297 is still there. This car and one or two of the wrecked steel flat cars are the sole surviving equipment of the logging railroad.
	TANK CARS: A total of six tank cars were owned. Most, if not all, were purchased from Tidewater Oil Company (Associated).
1-3	Used as water cars, and were equipped with diesel pumps for fire fighting.
DMCX 1-3	Three tank cars were used for oil cars. Two cars supplied Butte Meadows with oil from the sump at Stirling City. Principal use for the oil at the camp was for fueling the Shays. The cars were parked on an inclined track parallel and above the track the Shays were parked on. The third oil tank car stayed in Stirling City, where it was used to supply the sawmill boilers, on the east side of the log pond, with oil from the sump on the west (or town) side of the

*Approximate numbering in each series. Total number of each series and exact numbering is not known.

Series	Notes
	pond.
No Number	MOVING CAR: One moving car (i.e., heavy duty flat car) was owned for the purpose of moving the logging donkeys. Donkeys were not used for any purpose after 1938, but the car was retained for moving the cats. Sometime after 1938, the Stirling City car shop removed the deck from the car and installed log bunks. The car was then numbered in the flat car series, but the number is unknown.
No Number	SNOWPLOW: The snowplow was built by the Stirling City car shop just prior to the opening of the Butte Meadows line in 1935. It was approximately 28 feet long, with a V-plow mounted rigidly on the front. After the snowplow was in service, it was discovered the rigid V-plow had a disconcerting habit of piling up a ball of snow behind the plow while being pulled backwards, which would build up sufficiently to derail the plow. This difficulty was overcome by the installation of an air ram, enabling the plow to be raised. The snowplow was ballasted with scrap car wheels piled on the end of the car. Ahead of the car wheels, and behind the plow, was an open space where the section crews usually sat. Anyone sitting in this space was relatively protected from wind and snow. The snowplow was not self-propelled and had to be pushed by locomotives.
	SPEEDERS: A total of eight speeders were in service. All but two were built by the Stirling City shop.
Numbers Unknown	4-man speeder. At least three of these were used, all equipped with a one-cylinder engine. Assogned as follows: one to Stirling City yard; one to Harvey Breedlove (head electrician and plumber); and one to Jim Jones (Stirling City section foreman). Three larger speeders were built in the mid-1920's:
1	Buick 6 motor; eight-foot spread between axles; seated 12 to 14 men, with trailer seated 24 men. Built 1924-25 for steel gang.
2, 3	Dodge motor, six-foot spread between axles, seated 12 to 14 men. All three large speeders were used on the west side during the operation of the incline. A serious train-speeder collision during this time killed one man on the speeder. The speeder, either No. 2 or 3, was nearly demolished and subsequently was scrapped.
4, 5	4-man speeders. Fairmount speeders purchased in mid-1940's. All remaining speeders were scrapped in 1953.
	LOCOMOTIVE CRANES:
1	American 70 ton, purchased in 1923 or 1924. Fell through the Bull Creek trestle shortly after it was purchased, rebuilt and remained in service until abandonment. Used for scrapping operations until turned over by its operator in the middle of the Stirling City "junk yard." Scrapped on the spot.
2	American 90 ton (without boom), purchased from Utah Construction Company in 1936 for $9500. At the time it was located on a barge in San Francisco Bay at one of the construction sites of the Bay Bridge. Remained in service until abandonment, used by Commercial Construction Co. after they turned No. 1 over. Scrapped at Stirling City a few months later.
	LOGGING DONKEYS AND McGIFFERT SKIDDER: The logging donkeys and the skidder were numbered in the following series:
1-4	Williamette two-speed logging donkeys, which were first purchased in 1920. No. 3 was a wideface skyline rig.
5	Williamette three-drum, compound geared engine, which had been purchased in 1921. Used for the Coon Creek incline.
6-8	Winches or loading rigs.
9	McGiffert Log Loader and Skidder, believed purchased new in 1921. During the last years of the logging railroad, it remained in one location at Soda Springs, where it was used to reload logs from trucks to railroad cars. Scrapped by Commercial Construction Company at Soda Springs in 1953. The McGiffert was popularly known as the "jammer."

Wood, truss-rod flat No. 643 was typical of the early day log flats. This picture taken at Stirling City sometime between 1909 and 1915. (John Nopel collection)

Three types of cars are in the view above. No. 62, closest to the front, was a converted tank car. No. 37, coupled to it, was one of the old wood flats. No. 11, being loaded by Crane No. 2, was a skeleton flat. (John Hamby) At right, Crane No. 1 and oil tank car No. 1 near West Branch in the 1930's.

Old wood flat converted to ballast car, shown at Stirling City, April, 1950. (Robert Lowry)

"Maverick" high number flat car after World War II was No. 120, a rebuilt Russell car. Note the two air hoses — straight and automatic. Butte Meadows, September, 1949, with Conductor Charles Shelton standing on car, and Engineer Harold Beaver's in No. 106's cab. (Ted Wurm)

Harvey Breedlove, Sr., on a meter gauge speeder at Breakneck Trestle in 1915. (Bill Breedlove collection)

Speeder No. 3. From left: Roll Campbell, Mrs. Stanley Keesy, Mrs. Campbell. (Cliff Gilbert collection)

The snowplow at Stirling City, April, 1950. (Robert Lowry)

No. 297 was one of several retired refrigerator cars purchased from F. L. Botsford. (Stanley Borden)

Oil tank car DMCX No. 1 near West Branch in the 1930's. (Elmer Marion)

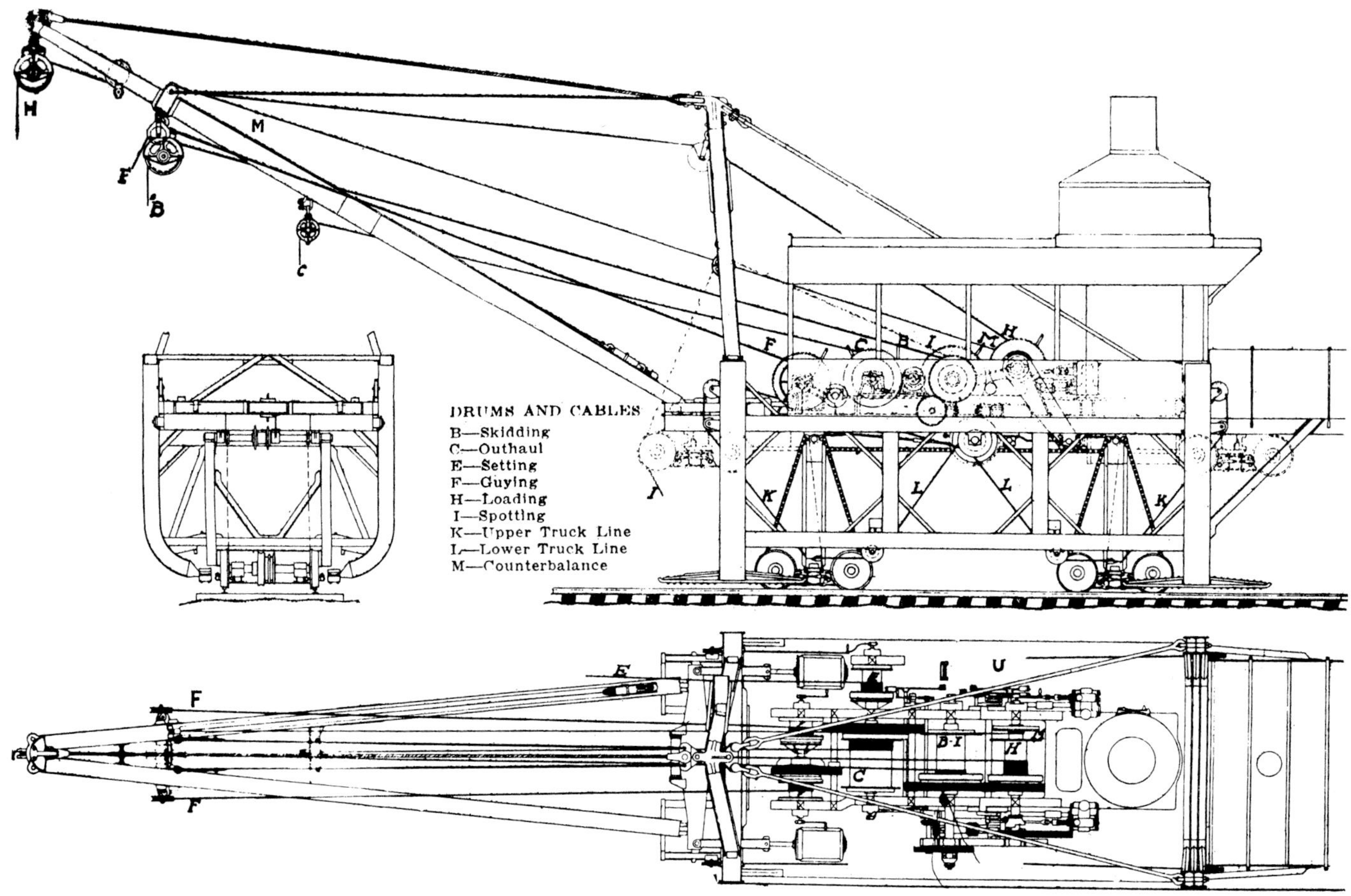

Diagram of a McGiffert loader similar to No. 9. (Reproduced from The Timberman, *courtesy of* Forest Industries)

FOOTNOTES

CHAPTER THREE

1. "Stirling to be summer resort." *Chico Daily Record* (January 14, 1904), p. 1.
2. W. H. Hutchinson, *The California Investment of The Diamond Match Company,* p. 285-286.
3. *Ibid.,* p. 76.

CHAPTER FIVE

1. *Ibid,* p. 24.

CHAPTER SIX

1. C. Raymond Clar, *California Government and Forestry—II: During the Young and Rolph Administrations,* p. 107.

CHAPTER SEVEN

1. Sylvia Cooke, "54-year old Stirling City mill closes." *Chico Enterprise-Record* (February 1, 1958), p. 2.
2. Sylvia Cooke, "Historic Stirling City mill destroyed by raging fire." *Chico Enterprise-Record* (April 10, 1959), p. 2.

CHAPTER EIGHT

1. W. H. Hutchinson, *The California Investment of The Diamond Match Company,* p. 53.
2. Clarence F. McIntosh, "Murder in Dogtown." Butte County Historical Society *Diggin's.* Vol. 3, No. 1, p. 9-10.
3. "Stirling City will soon be a dry town." *Chico Record* (December 5, 1907), p. 5.

CHAPTER NINE

1. "Burnie Broyles, passenger on caboose, tells of wild ride." *Chico Enterprise* (June 26, 1909), p. 4.
2. "Fail to recover body of missing train man Amos." *Chico Enterprise* (June 26, 1909), p. (1).

CHAPTER TEN

1. "Tells her experience." *Chico Daily Enterprise* (April 18, 1910), p. (1).
2. "Series of accidents happen at Stirling City." *Oroville Daily Register* (March 31, 1910), p. 4.

Bibliography

Books

Adams, Kramer. *Logging Railroads of the West.* Seattle, Superior Publishing Company, 1961.

Bryant, Ralph Clement. *Logging.* Second Edition Revised. New York, John Wiley & Sons, 1923.

Butte County Railroad Company. *Pine Chips.* Chico, Calif., 1904.

Diamond Match Company. *The Diamond Years. Commemorating the 75th Anniversary of the Diamond Match Company, 1881-1956.* New York, 1956.

Diamond Match Company. *Twenty Seven Years, 1881-1908.* Akron, Ohio, 1908.

Dunscomb, Guy L. *A Century of Southern Pacific Steam Locomotives.* Second Edition. Modesto, Calif., Modesto Printing Company, 1967.

Guinn, J. M. *History of the State of California and Biographical Record of the Sierras.* Chicago, The Chapman Publishing Company, 1906.

Holbrook, Stewart H. *The American Lumberjack.* Enlarged edition of *Holy Old Mackinaw.* New York, Collier Books, 1962.

Hutchinson, William H. *California Heritage. A History of Northern California Lumbering.* Chico, Calif., Diamond Gardner Corporation, 1957.

Hutchinson, William H. *The California Investment of The Diamond Match Company.* Chico, Calif., Diamond Match Company, Lumber Division, 1957.

Johnston, Hank. *Railroads of the Yosemite Valley.* Corona del Mar, Calif., Trans-Anglo Books, 1963.

Johnston, Hank. *Thunder in the Mountains: The Life and Times of Madera Sugar Pine.* Corona del Mar, Calif., Trans-Anglo Books, 1968.

Manchester, Herbert A. *The Diamond Match Company: A Century of Service, Progress, and of Growth, 1885-1935.* New York, The Diamond Match Company, 1935.

Mansfield George C. *History of Butte County, California.* Los Angeles, Historic Record Company, 1918.

McCulloch, Walter F. *Woods Words.* Portland, Oregon, The Oregon Historical Society and the Champoeg Press, 1958.

Wilson, Neill C. and Frank J. Taylor. *Southern Pacific: The Roaring Story of a Fighting Railroad.* New York, McGraw-Hill, 1952.

Periodicals and Newspapers

"Diamond Donkeys and Other Machines: Interesting Chapter in Diamond Match History" *The Timberman,* Vol. 52, No. 6, April 1951.

"Diamond Match Incline Real Engineering Feat" *The Timberman,* Vol. 30, No. 3, January 1929.

Dodge, Alexander W. "The Early Logger in the Sierras." *American Forestry,* Vol. 20, No. 11, November 1914.

Garver, John A. "Matches." *The Popular Science Monthly,* Vol. 11, 1877.

"A Great American Industrialist." *The Diamond Good Fellowship Bulletin.* Memorial William Armstrong Fairburn, 1876-1947. New York, 1947.

Johnson, Bob, and Jack Gibson. "1953 Report . . . Four Logging Railroads Go." *The Western Railroader,* Vol. 17, No. 4, February 1954.

Sagle, Lawrence W. "Lights! Camera! Action! Recollections of a Man Who Stabled Iron Horses for Hollywood." *Trains,* Vol. 22, No. 6, April 1962.

Periodicals and Newspapers

Chico Daily Record

Chico Enterprise

Chico Enterprise-Record

Chico Semi-Weekly Record

Chico Record

Diamond Digest

Diamond Good Fellowship Bulletin

Oroville Daily Register

Pacific Coast Wood and Iron

Pacific News

Pioneer Western Lumberman

The Timberman

The Western Railroader

Government Publications

Andrews, John B. *Phosphorus Poisoning in the Match Industry in the United States.* U. S. Bureau of Labor, Bulletin No. 86. Washington, 1910.

Unpublished Materials

Brouillette, Joseph F. *A History of Stirling City, Butte County, California.* Unpublished M. A. Thesis. Chico, Calif., Chico State College, 1959.

Hutchinson, William H. *Jay Gould and the Butte County Railroad.* Unpublished manuscript in possession of Diamond International Corp., Red Bluff, Calif.

Index

5